The Question of Limits

We have forgotten how to think about limits. Most philosophical approaches to the environment have focused primarily on the value of the natural world, the status of anthropocentrism and the Anthropocene, and the largely ethical questions of our impact on the world. While fully acknowledging these concerns, this book emphasizes the centrality of the confrontation between the imperative of growth that has been present since the Enlightenment and our belated rediscovery of limits. The expression "Limits to Growth," the title of a famous book from 1972 by Donella H. Meadows *et al.*, may have passed into a common discourse, yet the notion of limits itself remains insufficiently theorized, or even reflected upon, in the current movement of environmental advocacy. Sometimes it even seems as if there is an effort to avoid it.

This book argues that, on the contrary, we can only resolve the present global challenges by confronting the question of limits and making it central to our reflection. This entails discussing the long history of thinking about limits in which Malthus is the most infamous figure, but which also includes such major participants as John Stuart Mill and Karl Marx. Ultimately, *The Question of Limits* contends that the value of embracing limits extends beyond the environment and offers the potential to become a transformative social good.

The Question of Limits will be of great interest to students and scholars working at the intersection of environmental studies, economics, intellectual history and philosophy.

Christian Marouby is an Emeritus Professor of Mills College, USA. He is the author of two previous books, *Utopie et primitivisme: Essai sur l'imaginaire anthropologique à l'âge classique* (1990) and *L'économie de la nature: Essai sur l'anthropologie d'Adam Smith* (2004).

Routledge Studies in Sustainability

Towards the Ethics of a Green Future
The Theory and Practice of Human Rights for Future People
Edited by Marcus Düwell, Gerhard Bos and Naomi van Steenbergen

Sustainability Transitions in South Africa
Edited by Najma Mohamed

Strongly Sustainable Societies
Organising Human Activities on a Hot and Full Earth
Edited by Karl Johan Bonnedahl and Pasi Heikkurinen

Managing Complexity, Earth Systems and Strategies for the Future
Walter R. Erdelen and Jacques G. Richardson

Measuring Intangible Values
Rethinking How to Evaluate Socially Beneficial Actions
Marie Harder and Gemma Burford

Education for Sustainable Human and Environmental Systems
From Theory to Practice
Edited by Will Focht, Michael A. Reiter, Paul A. Barresi and Richard C. Smardon

The Politics of Sustainability in the Arctic
Reconfiguring Identity, Space, and Time
Edited by Ulrik Pram Gad and Jeppe Strandsbjerg

Global Planning Innovations for Urban Sustainability
Edited by Sébastien Darchen and Glen Searle

The Question of Limits
A Historical Perspective on the Environmental Crisis
Christian Marouby

www.routledge.com/Routledge-Studies-in-Sustainability/book-series/RSSTY

The Question of Limits

A Historical Perspective on the Environmental Crisis

Christian Marouby

Routledge
Taylor & Francis Group
LONDON AND NEW YORK

earthscan
from Routledge

First published 2019
by Routledge
2 Park Square, Milton Park, Abingdon, Oxon OX14 4RN

and by Routledge
52 Vanderbilt Avenue, New York, NY 10017

First issued in paperback 2020

Routledge is an imprint of the Taylor & Francis Group, an informa business

British Library Cataloguing-in-Publication Data
A catalogue record for this book is available from the British Library

Library of Congress Cataloging-in-Publication Data
A catalog record has been requested for this book

ISBN 13: 978-0-367-66461-9 (pbk)
ISBN 13: 978-1-138-33486-1 (hbk)

Typeset in Goudy
by Wearset Ltd, Boldon, Tyne and Wear

To Elizabeth

Contents

Preface

The question of entitlement

For reasons that will become clear if you read this book, I always had in mind that its title would be *The Final Stage*, with *Reflection on the Question of Limits* as a subtitle. The savvy editors at Routledge convinced me that it was too elusive – after all, it is not a book about the demise of the theater – and to turn it around, with a new subtitle indicating more clearly its historical dimension and the nature of the limits in question. Anyone can form an opinion about which was a better title. I am personally a little uncomfortable about calling it historical, since it is only so in the sense of intellectual history, and I especially regret losing the notion of stages, which is so central to my theoretical framework. But then, it is one of the arguments of the book that we cannot have everything.

It is however, the loss of another word that I want to briefly evoke for the moment. This book truly is a *reflection*. An attempt to look at the question of limits through its many facets and see what, and even how, it makes us think. Or perhaps an essay, in the original sense of trying to find a way to possible answers, to open a perspective, or a prospect, through the most important and challenging problem confronting our world. This doesn't mean that it is an objective, dispassionate reflection. On the contrary, as the previous sentence suggests, it is driven by a passionate engagement, a sense of urgency – even, as in the original meaning of "passion," a kind of suffering. But this attempt to reflect, rather than follow a predetermined path, explains in part the form of the book, which for lack of a better word can be thought of as organic. One chapter is distinctly longer than the others, because it needed to grow that way. Most of the three chapters that make up the middle of the book are much more about the past, even if not exactly in the way real historians look at it, while the opening and final chapters are much more about the present. The book is written to be accessible to any attentive reader, with no specialized terms left unexplained: in spite of my best intentions, however, it probably remains in large part an academic book, especially when engaging with other writings. So it is, in the end, a sort of hybrid, with its own stubborn integrity, because it is responding, as I hope my readers will also find, to a kind of necessity.

And then there is the other question of entitlement. Why me, you might ask? I am, after all, neither an environmentalist, nor trained as an economist, nor even a real historian, coming instead from the sprawling field of literature and language, my native French in particular, whose cause I served conscientiously, if not altogether faithfully, for many years. I could of course say, with some evidence to show for it, that in fact much of my work and commitment throughout that time had prepared me for this engagement, and that when I became free to write what I cared most about, it imposed itself, again, as a kind of necessity. And this would be true. But I think the more embracing answer is: why not us? Why not the many of us from all the fields concerned with how humans think and feel and express themselves, and create meaning from their private and collective experience, both now and in the past, from which we have inherited who we are? For the question of limits, the limits to our world, is ours to pose and to try and answer. All of us, regardless of our provenance and avocation. And so the subtitle of this book should really be: *A Humanities Perspective on the Environmental Crisis*, helping us understand that the crisis in question is above all a human predicament. Perhaps it would also make it more attractive to those who, while they care deeply about this question, are loath to read more specialized and scientific studies, worthy as they are of their attention.

To invoke the authority of a thinker whose importance for our reflection will soon be recognized, Adam Smith, who perhaps set a greater score than any other on the division of labor and its "peculiar trades," also recognized the importance of "philosophers or men of speculation, whose trade it is, not to do anything, but to observe everything; and who, upon than account, are capable of combining together the powers of the most distant and dissimilar objects" (Smith, 1776, I.i.9). In truth, those he had in mind were more inventors or engineers than philosophers in our modern specialized sense, but his point is still to be taken seriously. In our age of exquisitely fine division of labor, including that of intellectual labor, from which we have reaped such extraordinary progress in knowledge, it has become all the more necessary for us to take the time to "observe everything" that has resulted from this knowledge. It is also true that Smith's own time was probably the last when a single person of genius could still claim near mastery of all the contemporary fields of science and speculation. In ours, not only the level of mathematics necessary to understand the natural sciences, and often their social sisters, but even the highly specialized language of philosophy itself, have made this impossible. But it does not mean that we, the ordinary thoughtful humans, have to give up on knowledge, or indeed on science (which as we remember, means the same thing). We just have to trust those who have devoted their lives to particular fields of study, and speculation, and action, and have made the effort, directly or indirectly, to make available to us the fruits of their labors.

My gratitude, therefore, goes first to all the environmentalists, and ecologists, and brave economists I have never met, except through their writings, and to whom I owe almost everything I was able to think. For there is in truth little in this book that has not in some way been said before, so long already has the alarm of limits been eloquently sounded. I am thinking in particular of the pioneers who must have felt – some of whom alas must still feel – that they were crying aloud in the wilderness. It is for us to continue their calling, however many times and in however many ways as necessary, so it is finally heard. Of course, I have not remembered, or cited, everyone who has contributed to this reflection, and scholars in every relevant field will no doubt regret some omissions, but I hope most of those I admire will find themselves recognized. I am also aware that a great majority of my references are to studies and thinkers published in English, in large part because this book was conceived and written from my experience in the United States, where I feel the call still needs to be heard the most. This does not mean I do not recognize the contribution of all the work being done in other countries and languages, and in particular that of my other compatriots, whose presence is perhaps surprisingly discreet in these pages, even though I no doubt owe them, aside from the concept of *décroissance*, a good part of the way I think.

Much closer to my small world, I want to express my gratitude to Meléa Emunah, a real student of environmental science, for her invaluable help with factual research and for keeping me straight.

I also want to thank some of my close friends who have in some way contributed to the book: Alan the farmer, for showing us how to do it right; Dominique, for the Thursday conversations; and Dorothy (this is slightly more mixed) for suggesting long ago that "next time, you should write it in English."

And most of all, the dedicatee of this book, the *sage femme* who held my hand when I did not have the heart to carry on, and finally helped me push so it would come out into the world.

Reference

Smith, Adam (1776) *An Inquiry into the Nature and Causes of the Wealth of Nations*. Oxford: Oxford University Press (this edition 1976).

1 The question of limits

We have forgotten how to live within limits. Even as the world is rediscovering its finitude, we no longer seem to know how to *think* in terms of limits, let alone embrace what that might mean, even though we are increasingly aware that our universe is limited. Such is the paradox with which to begin.

But already we might question whether there is any paradox to begin with. After all, we must be living within some viable limit, otherwise we could not be living at all. At any given present time, this is so true as to be tautological. And indeed, there are still many millions of human beings in the world who are every day threatened with local or individual limits of the most basic means of sustenance. For them, and for countless others who are barely eking out their daily bread, or rice, or sorghum, there is no need to be reminded of the existence of limits. And so it is to a lesser extent even for most of us who are fortunate enough not to be in such dire straits: our ordinary world, on what we might call a micro level, is still governed by a sense of limits. This is the traditional, intuitive, common sense that always was the principle of sound husbandry, the rule of good housekeeping, the virtue of the Greek *oikos*.[1]

But over the last half-century or so, we have also become more and more aware, whether rich or poor, of the existence of critical limits affecting us (and being affected by us) on a wholly different scale, one that can by contrast be conceived as the macro level, and is appropriately referred to as global. Though these limits do not by any means affect us all equally, and some may long remain regional in their direct effects, they are ultimately the limits of the earth itself, the single home we all share and is therefore our common heritage: a finite and closed planetary system energized almost solely by the sun.[2] This emerging awareness of our common finitude, even if it echoes and in some ways recalls earlier, premodern forms of consciousness, is wholly unprecedented. It could potentially change our sense of being and the course of our becoming in the world, giving rise to sentiments of solidarity and joint responsibility without antecedent in the fragmented history of humanity – or alternatively, to equally unprecedented rivalry and mutual destruction if it cannot be jointly assumed by disparate nations.

This new consciousness of global limitations, however, also suffers at the outset from the very scale, indeed from the apparent indefiniteness and imperceptibility of the limits in question. Unlike the common sense of limits that is our traditional *habitus*, the notion that there are absolute limits to the world available to us, let alone that we could be pushing those limits to the point of endangering its survival, is to most of us only an abstraction, if it is imaginable at all. While both are of course the result of the same laws of nature whose patterns have long been observed by human sagacity, the latter are to a large extent only perceivable through scientific measurements, often made possible only by very recent technological developments, and even then, in many cases, with inevitable degrees of uncertainty.[3] They do not speak to our senses, and only indirectly to our imagination, except perhaps for those rare moments when we may be touched by an existential experience of finitude that could be called Pascalian (though Blaise Pascal himself, as a scientist and philosopher on the other edge of modernity, was moved to such consciousness by the mirror visualization of infinity).[4]

This first order of difficulty with the conception of limits is well known. It is the bane of all environmental scientists attempting to convince us that the most profound changes are rarely the most visible. Of course, the demise of krill in the oceans is more fundamental than that of polar bears on the ice floes, but we cannot help being more moved, and motivated, by the latter. It also explains why the most successful environmental campaigns to date have relied so heavily on visual imagery, and why some of the most influential environmental writings might in retrospect strike us more with their literary merit than their scientific value. It can also work the other way. Hurricane Katrina, and a number of other nearly contemporary natural catastrophes in different parts of the world, though largely coincidental to long-term trends, have probably done more to raise the consciousness of climate change than any rational argument to that effect.[5] Most famously, it is to one single photograph sent from the moon on the first Apollo mission that we owe the ability to actually see what no human being had ever been able to picture before: the miraculous beauty of our living planet so neatly defined against the surrounding emptiness. No wonder that it has become one of the most reproduced, and likely one of the most influential icons in the world, indeed the icon of our world. It provides us with the visual representation of our own finitude.[6]

But a greater obstacle to the conviction that we have entered the age of global limits lies in the other dimension of our perceptual subjectivity, that of temporality. The timescale of many of the phenomena involved is simply not commensurate with ordinary human experience, even over several generations, even over historical time, although they may be accelerating noticeably. This calls not only for further reliance on highly technical data, sometimes going back to eras when nothing like human life had yet appeared on earth, but more crucially, on projection into the future. For the fact that the limits we live within must be viable at the present time (in the tautological sense evoked above) does not mean that the cumulative effects of our way of life will permit this to continue indefinitely. They may well be on a trend to overwhelm the limits of viability at some future time, not only those of our present lifestyle, but of any life at all. But how soon, or how far into the future?[7] Most people naturally extend their sense of concern and responsibility to the lifespan of their children, and of course to that of their grandchildren. And surely, to the next generation ... In principle, we care indefinitely about the future fate of humanity, but we do so more and more abstractly as we project further in time, so that it is hard to know how much such a sentiment really counts beyond a few generations, even if most people were persuaded that a human impasse would eventually be reached.[8] This is the subjective and moral difficulty with the timescale of limits, and it is easy to recognize what formidable problems it poses, in particular for a political process based on short-term cycles. It explains in large part why environmental limits are often dealt with only in crisis mode, that is to say when they have already been breached, when it is already too late.

But even if we could count on perfect clairvoyance and an infinite sense of multi-generational responsibility, this still would not guarantee a universal conviction that a trend presently leading beyond viable limits at some point in the future will inevitably do so by the time we get there. Clearly this will only happen if none of the factors foreseeably affecting the trend changes, and if no new unforeseen factor appears in the meantime. But all retrospective historical experience teaches us on the contrary that unforeseen change will certainly occur, and even that some wholly unpredictable discovery or development is likely to alter the course of events in ways we cannot possibly imagine. And the further we look into the future, the greater the likelihood of such an event. Thus, as we introduce the dimension of time into our reflection, as of course we must, we find that the difficulty with the acceptance of limits cannot be reduced to issues of perception, or lack of imagination, or even of collective responsibility – although those are all very real obstacles – but that it also rests on a fundamental element of unpredictability at the heart of the question of limits itself. This does not mean, as is sometimes claimed, that present projections of future disaster can be ignored because human ingenuity will find solutions to everything – after all, things could also turn out much worse than anticipated – but it does require from those warning us of the imminence of limits the greatest caution. Earlier prophets of doom from Malthus to Paul Ehrlich,[9] no matter how right they might have been on theoretical grounds, have sometimes been famously wrong in their predictions, arguably doing more long-term harm than good to the cause they intended to serve. For this reason alone – even if it is not the only one, nor perhaps as we shall see the most substantial – it is understandable that the notion of impending limits should for a long time have been met with great suspicion, if not with outright denial.

Still, for all the obstacles and points of resistance acknowledged so far, it is undeniable that a new consciousness of global limits has been emerging and spreading rather dramatically during the past decades, concerning a wide range of variously critical environmental factors, some of which, most prominently global warming, have reached spectacular levels of world-wide attention and official recognition. Many others are so well known that they will come to mind immediately, and it is probably as unnecessary as it would be impossible to provide a comprehensive list here, even less to attempt some sort of consensual ranking by degree of urgency.[10] This, moreover, not only for the obvious reason that such data are large and constantly changing, and therefore better made available through direct and updated sources, but also for another reason that will require our full engagement at a later point, but must already be apparent: while some limits are "hard" in the sense of objective, and recognized as such by nearly everyone, many others, more subjective and in this sense "softer," are dependent on how much particular goods are valued and prioritized by different people or cultures, if they are recognized at all. Given this moral relativity of limits (which is conceptually of a completely different nature from the unpredictability identified above), establishing the full scope of what counts as such – even before we consider what can and should be done about them – would therefore constitute an infinitely complex and contentious task, as considering even a few familiar examples will show. Still, a sampling of some prominent cases might not only be useful in clarifying what is concretely at stake, but also give us a chance to assess how closely and how urgently we would place different goods on the threshold of limits.

Limits in the sense that we are concerned with here can be reached at the two ends of the flow of human interaction with the natural environment, those commonly designated by ecologists as "sources" and "sinks." The first limits are reached by exhaustion, when we can no longer extract from the earth a particular good we want to consume or use. The second are reached by overwhelming the earth's capacity to absorb the by-products of our use. In the category of "sources," those that from our instrumental perspective are viewed as resources, minerals and fossil fuels figure prominently, oil being the most talked about, with widely varying estimates of predicted exhaustion. All fossil fuels are similarly time-limited, albeit in different ways, and it is worth noting that uranium, the fuel for the energy that is often promoted as an alternative to fossils, is also finite, and in fact extractable only from very few places on earth. Gold is of course precious to us, but we could probably do without it. On the other hand water, the good that for classical economists was emblematic of a commodity without value because it was infinitely available, may well turn out to be the most crucial limit for human development, even if it remains globally abundant.[11] Forests are also limited only regionally as a resource, but they are of global concern for their function as a sink, and they are shrinking.[12] Oceans and seas, we imagine, cannot disappear, but their contents can. We are already consuming more "farmed" seafood, that is to say industrially bred in ponds and enclosures, than fish from the sea, because their numbers are dwindling.[13] And that, after whales and polar bears, is only the most visible loss. Beyond krill, there are thousands of species of microorganisms in the oceans alone that are becoming extinct before they are even known. Rates of extinction on land are also much faster than would naturally occur, including those of plants that may potentially have proved invaluable to us or to the balance of the overall ecology: we will never know.[14] In such cases the threshold of limits is not in some probable future, it is now, and already yesterday.

And then there are those creatures or natural landscapes that are not necessarily "useful," but are beautiful and magical to us. And there are those that are not beautiful to us, or simply invisible, but just as essential an outcome of evolution, and a part of the long-term equilibrium of the earth. But even without calling into question our anthropocentrism, what will our descendants think when they learn that we finally removed from their world, deprived of their natural existence, the last descendants of our own near ancestors, our genetic cousins the great apes? Not a crime against humanity, exactly. But surely there are already many of us who feel that by crossing such limits, and many others, we are disgracing our responsibility as reluctant custodians of creation (even if we don't believe in any intentionality in the universe). Others, fleeing from actual war crimes, if not themselves half-coerced perpetrators, may well think of such exalted sentiments as the moral luxuries of the West.

There is hopefully no need for a similar litany on the side of "sinks," some of which have already been mentioned as threatened sources. With the complementary concept of pollution, they were the first to come to the attention of the environmental movement, if we think of such an epoch-making classic as Rachel Carson's *Silent Spring* (1962). Regionally, as a result of this organized environmentalism in the developed world, the measurable condition of sinks such as the air and water and forests has actually improved since that time in a number of mostly Western countries. As we will have to acknowledge in Chapter 5, however, this has been achieved in large part because those rich nations have massively exported their dirtiest industrial work to poor developing ones, in addition to actually dumping some of their own refuse in the most desperate foreign sinks.[15]

Morally, the question of environmental damage is not radically different or always distinguishable from that of the depletion of resources, in that both tend to spread misery and exacerbate regional inequalities and injustice. Economically, though, we will find that sources and sinks present very different problems and opportunities for equitable solutions, in that scarce resources, at least in principle, respond to market mechanisms and incentives by becoming more valuable, whereas the overflow of sinks, being less than worthless, is a burden no one wants to take responsibility for and becomes a pure collective cost. This is particularly evident when the "dump," not being located in any one's particular backyard, is in everyone's backyard, that is to say when it has become global, and growing. And it is precisely in this last category that we find the most famous example of a failing sink on a worldwide scale, global warming. Because it is so well known, so potentially consequential, and because it has mobilized unprecedented media and political attention, it presents an exemplary opportunity to test the question we have been opening.

Global warming may be as clear a case of limits as we will find. Here we have an objective, physical, measurable threshold, that of the temperatures within which life is tolerable, or conversely beyond which human civilization could not endure without extreme losses and hardships. There is perhaps a range of survivable variation and adaptation with massive migrations, but it is small. We have, by now, a generally agreed upon cause: the excess accumulation of greenhouse gases in the atmosphere because more are generated on an annual basis than can be absorbed by the natural sinks, essentially the biomass of plants on land and in the oceans. There is some disagreement over what percentage of this excess generation is caused by human activity in relation to what the natural environment would spontaneously produce at this moment of climatic history. But the consensus is also growing, and in any case there is no doubt as to which human-induced processes contribute to the excess: primarily the combustion of fossil fuels for the production of energy and heat, followed closely by the destruction of the world's primeval forests, and secondarily (this is not funny) the extra methane released by cattle bred for human consumption.[16] And so, most remarkably, we know perfectly well what the solution requires: to reduce our production of the same greenhouse gases (or alternatively or in combination, to proportionally increase the natural or artificial absorption of those gases). This does not mean that we understand all the factors and complexities of climate change, or that we can predict with certainty the chain of consequences that would be unleashed if global temperatures rose beyond their historical range. But we do recognize this as a dangerous and non-negotiable limit, we know that it is being threatened, and for the modest part that is ours to control, we know what to do about it.

In fact, we even ran a kind of pilot program for it in the 1980s, in what is considered (by those who remember) a model of relatively if not perfectly successful environmental crisis resolution: the breach and repair of the ozone layer. Then too we had a global threat (though the holes were local in place and time) caused by an excess release of industrial gases – in that case the much rarer chlorofluorocarbons – and an appropriate remedy, an international agreement to ban the use of said gases (the Montreal Protocol of 1987), which was ratified and successfully implemented in a relatively short time.[17] Of course, the scale and complexity were of different orders of magnitude, which would certainly account for the much greater difficulty in making significant progress to implement efficient measures against global warming. But one senses that the issue of scale is not the only reason for the difference, nor even the principal one.

Whatever it is, the reason for our failing to act responsibly and appropriately against the threat of global warming – a prime example of what it would mean to live within limits – is no longer the lack of recognition of the problem or attention to it. Unlike as recently as three or four decades ago, when the idea that we could be reaching limits was still greeted with general derision or simply dismissed (with notable exceptions), it is now acknowledged – at least in the case of this one limit, and as it were thanks to it – publicly and officially. And this, it is worth reminding ourselves, at the highest scientific, organizational and symbolic levels the world has yet achieved. No less than 23 (and counting) special United Nations sponsored conferences (COP) were devoted to it, leading to the adoption of several international protocols and an ongoing process of negotiations and ratification.[18] Just as memorably, a Nobel Price for peace was awarded to a former American Vice President for helping bring attention to it through a well-publicized media campaign and film.[19] The global warming crisis could one day be remembered as the discovery of the first limit –

a concept historically symmetrical to, and therefore mythically resonant with that of the last frontier. Though it may not yet be interpreted in quite these terms, that is what it means: the first official recognition of a global limit.

It is all the more striking, then, that during the more than two decades since the crisis was first internationally recognized and appropriate measures adopted in principle (Rio Earth Summit, 1992), the worldwide emission of the greenhouse gases clearly identified as the cause of global warming has not only not been reduced, nor even stabilized, but has on the contrary kept increasing substantially, and is on a curve to grow even more in the realistic future.[20] Since it spotted the proverbial iceberg, to use the well-worn allegory, the *Titanic* has not only kept its course, it has continued to turn on the gas. Now it is true that for all its rhetorical appeal, we should avoid the provocative irony of such a paradoxical statement. Given the immense problem of "turning around" global warming, it is not surprising that finding and implementing commensurate solutions would take time and meet with difficulties, in particular political difficulties, in a world divided by such monstrous inequality. This last point especially will not be ignored, since the issue of equity is in the end a fundamental horizon of the question of limits.[21]

But for the time being it is to a more practical difficulty we must first turn: what *kind* of solution are we looking to implement? And it is from this angle that the comparison with the ozone layer crisis may be most useful. In that case, a technical solution, finding substitutes for the offending gases, was relatively easy. Through these other means, we could keep making just as many, or more, refrigerators and aerosols as before, without our life being substantially affected by the change.[22] We hardly noticed. To a very large extent, it is the same approach that is unconsciously guiding our attempt to solve the global warming crisis: to find technical substitutes that would not require us to change our way of life, and allow us to continue consuming, and producing, and traveling just as much as before. Only this time it cannot work, because the available substitutes, the truly "alternative" sources of energy that could indeed save us from our sinking problem, are of a different nature from the ones that have sustained our recent burst of activity. They have a different kind of finitude.

The crucial difference in this case is that between flows and stocks. Flows in this sense are all the energy processes active on the earth at any given time, such as winds and rains and tides, and the whole life of the biomass. They are all ultimately driven by the sun, with the exception of those resulting from the residual inner heat of the planet. Flows were the only sources of energy humans knew how to use (if we count, as is normally done, wood from renewable forests) until the industrial revolution. Stocks are the accumulated, mineralized (hence fossil fuels) products of millions of years of organic processes; they are our providential inheritance from the untapped flows of all the millennia of life on earth.[23] Bio-power from the past. But in just the last 200 years – and even then, most massively in the last quarter of that short period – humans (and until recently, only a minority of us) have learned how to take advantage of this bounty to light up a bonfire of increasing intensity, stoking it with ever more miraculous fuels as they were being discovered, as if there were no tomorrow. And indeed, it seemed that the party could go on forever, so much was there of this buried treasure. Now, suddenly, what would have been utterly unimaginable even a century ago has happened. In this fraction of the history of life on earth, we have managed to burn perhaps half of the stock, and the rest is going fast, or is getting more expensive and destructive to extract.[24] And as for the most abundant of what is left, the coal that started it all and could go on burning the longest, we are finding that it is also the worst offender of the greenhouse gas-producing fuels, with oil not so very far behind.[25] One way or another, we will soon have used up our inheritance of stock, or won't be able to draw on what is left.[26] We must now return to flows.

The solution, then (with the most prominent exception of nuclear power, a non-carbon-emitting form of energy with its own problems), is in fundamental ways to go back to the sources of energy that preceded the stock-based industrial revolution. It will of course not be a return to pre-industrial technologies (even if a wind turbine is still a windmill), benefitting instead from all the advances and efficiencies offered by modern science, including unprecedented possibilities to tap into tidal or geothermal sources, to collect directly the energy from the sun, and others we may not yet imagine.[27] But whatever improvements are added to our ability to harvest the energy of flows, it will have to respect their basic limitations, which are in many ways the opposite of those of stocks. Flows as sources of energy are almost indefinitely available, since they will not run out as long as the dynamic balance of the earth is not threatened, but they are strictly limited at any given time. More or less can be collected, depending on our total capacity, and some of it stored temporarily, as has long been done in hydroelectric reservoirs, but never more than the earth itself makes available in the present through its interaction with the sun's energy.[28] Indefinite but limited at a given time, instead of (almost) unlimited but finite.[29] Such is the trade off, and in the long run it is a very good one, if we can make the transition without too much pain. A human ecology based on flows, accepting as it must be of their slower pace, less destructive and frenzied than our current slash and burn policy, could also be wiser for us. If we can agree to live within its limits.

Confronted with this dilemma – and we have seen that it is beginning to be recognized, at least in some of its principles – the world has so far been spectacularly reluctant to choose, or has made only characteristically ambivalent commitments. On the one hand, as already noted, it has not by any real means given up on fossil fuels, despite increasing attention to global warming. While no-one would dispute that some continued use of stocks is indispensable to achieve a tolerable transition, and realistically, we probably should not expect their exploitation to end until it has become prohibitively expensive, the currently undiminished reliance on such knowingly destructive forms of energy suggests that something is not registering, and amounts to a practical denial of this first limit. On the other hand (still with the nuclear option being held in reserve, as a potential trump card) renewable energy sources are now being enthusiastically promoted by policy-makers and even invested in by innovative entrepreneurs, but in a mode that is suggestive of another kind of denial: it embraces their indefinite potential, without accepting the consequences of their limitedness. In this new age of the greening of all things, the development of alternative energies is seen as the new frontier of technological innovation, the next growth industry, indeed as the basis for the second or third industrial revolution.[30]

This may well be true, and all our scientific genius and inventiveness will not be too much to help us through the coming crisis. But wind turbines and photovoltaic panels and hydrogen cells do not grow out of thin air any more than computers and internet routers. They have their own costs in resources and labor and energy, and cause their own nuisances and toxicities. More fundamentally, renewable energy technologies do not abolish the laws of thermodynamics, or the principle of entropy, which put simply, says that we can never produce (put out) more than we receive (put in). If anything, the transition from high energy-intensive stocks to low intensity flows means that we must learn to use altogether less. But that is precisely the inconvenient truth[31] the green industry movement is trying to grow away from. It is highly revealing in this regard that the only way the goal of actually consuming less energy seems to be promotable, under the more neutral term of conservation, is as an opportunity for the insulation business in retrofitting old buildings, or building more efficient ones, and therefore for growth in the construction industry.[32] In this mindset, the solution to too much always turns out to be more.

But even if enough windmills could be built to crest every hilltop and enough solar panels to cover every desert and enough turbines to dam every river (or more likely, enough nuclear reactors) so that as much power as before were produced without pushing the threshold of climate change – if we developed ourselves out of this limit – then we would still be bumping against all the other limits we are currently threatening, and soon against many more as we thrive on our green energy. For we are not yet living in green houses, or shopping in green malls, or watching green screens fed by green satellites. For that matter we are not drinking green water, and we are not even eating green food, not to mention fighting green wars.[33] No doubt technological solutions can be sought to all but the last of these so that we can build and communicate and certainly eat with less pressure on precious resources, and less collateral damage to the environment. All are good and necessary goals to pursue and support, and their adoption by increasing segments of the public and private sectors is a hopeful sign for the future. The argument therefore is not against the greening of development, or even against green growth if it is possible, but for addressing the question that is thus being begged. What the dominant approach to the confrontation with limits reveals is a powerful impulse to avoid the question of limits itself, a refusal to face our condition as one of limitedness, not just on a practical level, but as our fundamental mode of being in the world. It amounts to an ideological denial of finitude.

This denegation is not coincidental, or a simple case of obliviousness on the part of the contemporary world. Modernity in its entirety, the movement of intellectual and social transformation that started with the scientific rebirth of the West and gradually conquered the greater part of humanity, can well be thought of as a refusal of limitedness, if not as a revolt against finitude.[34] This does not mean the rejection of limits themselves, since rationality and science are on the contrary conditioned by the acceptance of objective limits outside of our powers, by the disenchantment of our relation to the world.[35] But as will and intentional project, modernity has from the beginning been bent on pushing out and forward the limits of knowledge and human possibility.

This is true of science and technology, whose limitless quest has only been expanding and accelerating since the naively satiated triumphalism of the high Enlightenment. It is even more fundamental for the other legacy of modernity with which any reflection on limits must engage most deeply, the discipline that seeks to understand and theorize the material progress of society that we have come to know for short as economics, and the economic system itself as it has come to dominate the planet, diversely known as liberalism, free enterprise, or market capitalism. On the one hand, the notion of limits is foundational for economic thought, since there is no (economic) value, therefore no economic object or interest, without scarcity, that is to say some limitedness of resources. Indeed, some of the most compelling contributions of economic theory, such as the law of diminishing returns, have to do entirely with the recognition of limits. On the other hand, political economy has from its inception been motivated by the desire to expand those limits. Already in its adumbration in the first modern treatises on wealth in the sixteenth and seventeenth centuries, it has everything to do with the practical need to increase the riches of the Sovereign, and this goal becomes if anything more of the essence in the great founding theories of the late Enlightenment. In advocating for the removal of all barriers to trade and enterprise, for the "system of perfect liberty," the liberalist revolution only aims at the greatest possible expansion of the wealth of nations, in everything but name what we now call growth. Economic thought is as it were genetically committed to the overcoming of limits.[36]

No wonder then that ecology and economics, etymologically so akin, are instinctively so adverse. While as objective sciences both devoted to the study of the fundamental ways in which humans live in and from their "home," economics and ecology are in principle not only compatible, but could in some reconciled future be thought as one discipline, it is easy to see how their original and by now almost unconscious orientation with regards to limits propels them in opposite directions. Ecology, born out of protective allegiance to the natural environment, insists above all that we respect its unchanging limits. Economics, resolutely anthropocentric, is designed to free us from those limits in order to achieve infinite progress.

It is also true, as is often observed, that economics has long drifted from its etymological roots in the Greek home, or any home at all, to the point of betraying its original meaning.[37] If the poetic extension of the word throughout the early modern period from the "rule of the house" to that of the great household of the *polis*, resulting in our political economy, is both natural and understandable, it also masks the profound change the concept of "rule" itself has undergone in the process. For if the rule of good housekeeping is to live within limits, as is still retained in the domestic need to "economize" and the private virtue of economy, it would be a very bad economist who ruled the modern *polis* like a home by merely economizing. As only those not trained in economic science can perhaps still appreciate, the logic of capitalism is radically counter-intuitive, in that it defies our common sense about limits. It is a world in which spending more can make you richer; where increasing one's wealth need not decrease the wealth of another, but may in fact be mutually beneficial; and where profligacy is encouraged as a virtue. According to its alchemy, money multiplies the more it changes hands, and wealth breeds wealth indefinitely, instead of running out as it is expended. While there is nothing mysterious about this magic, it is well to remember that it was wholly unknown to Antiquity, and to the most opulent of all the great civilizations that rose and fell before us. Capitalism, as distinguished from the reliance on some form of accumulated capital that is the originary condition of all human progress, is as an economic concept and system unique to modernity, and together with science its defining development. In the long perspective of the human relationship with the natural world, its most outstanding characteristic is its defiance of what we might retrospectively call, in reference to the traditional, pre-modern view that preceded it, the rule of limits.

The only thing that might seem even more astonishing is that it has worked so spectacularly well. For in the same millennial perspective that we should try to keep in mind, the degree of material progress achieved by economic capitalism in the last 200 years is simply out of scale compared with any previous historical period even ten times longer. This is undeniable regardless of all the qualifications and reservations that can legitimately be voiced about this success. Of course, it is not only an economic accomplishment. The exponential expansion of productive activity that started with the Industrial Revolution could not have been achieved under any economic system without the scientific revolution that had been laying its foundation for close to 200 years, so that we should perhaps think of it more accurately as a techno-economic phenomenon. On the social, religious, or political conditions that may have facilitated the rise of capitalism (or conversely for some, that may have been its superstructural manifestation), opinions have long been famously divided, as they remain about the all-important question of the relationship between economic liberalism and democracy. Whatever causal theory or degree of determinism one subscribes to, however, there is hardly any doubt that the economic revolution of modernity is deeply connected with its larger social transformations.[38]

On the darkest side, neither can there be any doubt that its beginnings are contemporary with those of Western colonialism, and the period of its greatest expansion in the West with almost worldwide imperialism. For all but the last half-century or so, that great economic engine has functioned as a centripetal force for the overwhelming benefit of Europeans in their various centers of power. This, together with the memory of the brutal beginnings of industrial capitalism in Europe itself, is not only a moral blight on the reputation of liberal economics, but has real consequences that may prove ineradicable. In spite of the stupendous economic catching up under way in several formerly dominated nations in the last 30 years, or perhaps as a result of it, the levels of worldwide inequality created by the uneven distribution of wealth in the past centuries will likely be the most formidable obstacle to a humane resolution of the ecological crisis.[39]

For all that, it would be a grave mistake to underestimate the power of capitalism, particularly its ability to mobilize innovative and productive energy, and its formidable multiplier effect. The most famous and relentless of its critics, Karl Marx himself, had no doubt about this awesome generative capacity, and much of his immense work can be read as a reluctant homage to it. From the same perspective, it is also appropriate to remember that for all their differences in allocative and distributive priorities, and whatever one thinks of their merits and efficiencies, all the various socialist experiments that were ever attempted were as fully committed to economic and material growth as capitalist societies, and relied just as much on capital investment for that same purpose. If anything, they were even more ideologically resistant to the possibility of limits to human expansion and the appropriation of nature. In any case, it is by now a fact that practically all the emerging economies, including those from the former communist sphere, have finally embraced market capitalism with a vengeance, with or without democracy, precisely for its extraordinary productive capacity. Among these are the countries that have achieved the most phenomenal growth rates in the last 30 years and are so far continuing to outpace the older economies, so that regardless of their origins and history, the model and achievements of liberal market capitalism can no longer be considered a Western privilege or monopoly. In spite of huge initial inequalities, and while billions are still ignominiously left behind, a great if slow movement of re-equalization is now under way in some of the most densely populated parts of the world.[40]

This is a development of profound significance and hopefulness, and of course few liberal economists need to be persuaded of it – if anything it is too often used by apologists of liberalism to justify its abuses – but it is not sufficiently acknowledged by the other side. Ecologists and environmentalists, from whom this reflection draws much of its inspiration, and especially those coming from the Left, such as the anti-globalists or the better named (in the francophone world) *altermondialistes*, tend to be overly swayed by their principled aversion to the truly offensive aspects of capitalism, and thereby partly blinded to its beneficial effects and its powerful appeal for the wretched of the earth. When we think of the last 200 years of world history, particularly the last century, we are assaulted by the visible series of crises, depressions, world wars, genocides and fanatical destructions, and the persistent predations and famines. These are all too real, but if we look behind these events, and in spite of them, at the no less real economic history, what we see is a steady rise in living, educational, health and aspirational standards such as the world had never known, and this while overall population has also grown exponentially.[41] Yes, it is scandalously uneven and immorally distributed, but if the same increase in material prosperity could possibly be sustained and continued indefinitely at anything like the same pace, even with its egregious injustices, we should all swallow our moral reservations and root for it.

But it can't. It cannot continue in its present form for all the reasons outlined above, and above all because this extraordinary expansion was only made possible by the dilapidation of exhaustible fossil fuels and natural resources, with commensurate degradation of the environment and sinking of the world. We have therefore entered into an era of direct conflict between the goal of increasing material productivity for our benefit and the need to safeguard the imperiled planet that is our sole source of wellbeing. Between economics, for short, and ecology.

This conflict, it should be pointed out, is not exactly with "growth," at least not as it is normally defined economically, and in this sense the movements of social and environmental activism that are labeled "anti-growth," or in the more felicitous French term, advocating *décroissance*, are only useful as provocative approximations. For the indexes that currently measure economic growth, the GDP or GNP, are extraordinarily crude instruments to assess the environmental impact of human activity.[42] For one thing, since they count only products and services for which a price has been paid and in some official market, there is a lot that they leave out: illegal deforestation, for instance, would probably not count, and even less the dumping of polluting waste products, but neither would the environmental clean-up efforts of thousands of volunteers after an oil spill. Among the activities they do register, on the other hand, cutting a tree counts in the same way as planting a tree; the sale price of a Rembrandt as the revenue from a Renault plant (but not its CO_2 emissions), and that oil spill only shows up in the paid clean-up costs, and the fees of the lawyers, both as positive contributions. Without looking more precisely into the nature of each of the activities that contribute to economic growth, the overall measure of the "gross product" of any nation is much too gross to tell us directly about its impact on the environment and the depletion of natural resources, or for that matter on how much it actually improves the lives of its inhabitants.

To attempt to remedy this crucial lack, a number of other indexes have been developed and experimented with in the last decades. The most promising and relevant to our question seeks to measure the "ecological footprint" of individuals or any economic sub-group or country, the real impact of their lifestyles on the earth (Global Footprint Network, 2018). Rather predictably, it tends to confirm the suspicion that there are ecological Yetis (aka Big Foot) among us, but it can also reveal surprising differences between seemingly comparable standards of living. For this last reason, other, more comprehensive indexes have tried to measure economic "welfare" (sustainable or not), i.e. how well we really fare in a particular economy, a notion that is obviously much more elusive than the simple measure of economic transactions.[43] Neither of these approaches is without its own difficulties or beyond dispute, but together they suggest a kind of utilitarian formula that would surely generate a very large consensus as an objective for humanity: to produce the most welfare for the most people with the least overall footprint. What is striking about this way of measuring how well we are doing within our earthly home is that it doesn't say anything about economic growth, at least not directly.

Nevertheless, in the aggregate, and certainly as long as we continue to rely on elemental resources in anything like the way we do now, there is a very high degree of correlation, even if it is a rough one, between GDP and footprint, therefore between growth and environmental degradation. There is a conflict between the fundamental economic assumption that output should continue to increase indefinitely and the need to go back to living within limits. This is the really inconvenient truth we must face, though it is still met by most people with varying degrees of denial, even among those who do recognize the threat to the planet and are committed environmentalists. There are of course many reasons why one would try to convince oneself or others that the conflict will resolve itself without having to restrain growth or, Earth forbid, to deliberately engage in *décroissance*. Any public figure who presently advocated such a policy would surely be committing instant political suicide, even if there are more of us than one might think who would be willing to face the truth and its consequences.

But the denial need not be disingenuous. All the dominant models currently proposed to help us address the conflict encourage us to avoid it, primarily through the promise of technological progress. As we will better be able to recognize later after reflecting on the primacy and influence of the paradigm of stadial evolution, the most compelling of these arguments – in part because it contains an important element of truth – is what could be called the fallacy of sectoral progression.[44] As modern economies get richer and more advanced, they tend to develop from their original base in agricultural and raw material production (primary), and then industrial transformation (secondary), to finally emphasize the tertiary sector of service activities, many of which are more "virtual" in the sense of less material, and therefore result in a lesser ecological footprint per unit of wealth produced than do the heavier activities of the primary and secondary sectors. This is the part that is true. But if the portion of the economy with a relatively lighter footprint becomes proportionally larger, that does not mean that the heavier industrial activities sector has been diminished in absolute terms. On the contrary, it has continued to increase even in the case of the most advanced, most "virtual" economies, even as those have partly unburdened themselves by exporting the dirtiest and heaviest tasks to the emerging world. To convince ourselves of this, we only need to look at the massive levels of resource use and material consumption in the very countries that have developed the largest service economies. As *producers* of wealth, financial traders or software developers (or whoever gets the commission for that Rembrandt sale) may well be engaged in largely virtual activities with a low ecological impact, but as *consumers* – though this is undoubtedly a generalization and therefore unfair to some of them – they are hardly known to tread lightly on the earth.[45] Just as there is no software without hardware, there is no virtual economy without a real one.

More fundamentally, we must rid ourselves of the replacement model, that is to say of the illusion that more advanced modes of production take the place of earlier ones by a process of substitution. Rather, they speed up and multiply the productive capacity of the previously dominant sector by transforming it in their own image. Thus, the advent of industry did not eliminate agriculture, it industrialized it by mechanization and chemical fertilization, making it in the process vastly more efficient, but at the price of subjecting a formerly renewable mode of production to a complete dependence on fossil fuels. More recently, we can observe with our own eyes that the spectacular development of digital technologies, the much-heralded new information age, has not by any means superseded industrial production, but is in the process of computerizing it, again with unquestionable gains in efficiency and productivity, all resulting not in the replacement but in the multiplication of industrial output. There are now obvious signs, and a real danger, that the development of greening technology could be interpreted and made to operate in just this same mode, with cleaner and more energy efficient inventions promoted as the means for yet another wave of growth, a new opportunity to produce more, a different way to continue pushing against limits.

The green revolution, if it is to live up to its name, must on the contrary break with the fundamental logic of growth. It must not only use less (to produce more), but aim to produce less, or more accurately, to consume less, if we remember the original meaning of the word. This requires a profound change in the habits of thought that underlie economics, indeed in what has now become our horizon of expectations. And in this necessary and difficult enterprise, those who are most convinced of the productive power of capitalism and of its historical achievements should also be troubled by the possibility that this very power may also have become a liability. For it is not obvious that liberal economics, for all its inventiveness and ability to transform itself, can do away with the imperative of more,[46] so much is this principle ingrained in its very fabric, and as we shall see when we examine its genesis, inherent in its primary motivation.

These grave misgivings, it should be noted, while they may or may not be compatible with the Marxist critique of capitalism, are different from it, and stem from a different conflict than the one identified at the core of Marxism. What is striking about the present, perhaps transitory moment, is our apparent inability to recognize this newer conflict as such. In all the affluent parts of the modern world at least, citizens are now concurrently encouraged, often by the same public and private institutions, both to be environmentally virtuous, to conserve and recycle, to adopt green lifestyles, and at the same time to be good economic agents by creating as much wealth as possible and submitting to the relentless incitement to consume equally as much – as if there were not at least some tension between those two goals, not to speak of contradiction. And even when a possible conflict at last appears to be perceived, if not fully acknowledged, it is in the form of a kind of sublation that all but conjures it away.[47]

The most famous example of such a conceptual operation is the nearly universal embrace of the Brundtland Commission sponsored principle of "sustainable development."[48] Sustainable development is a good principle that should of course be supported, and whose United Nations adoption constitutes a huge step in the recognition of an ecological imperative. The concept of sustainability implies the existence of limits, and its implementation would amount to reducing (since we are already in excess) the total footprint of humanity to the carrying capacity of the earth. It raises many difficult questions of measurement and subjective choices, but no more than the notion of limits itself. Development also points to a significant corrective: to live within limits does not mean to live in a static, unchanging universe. It requires just as much inventiveness and technological innovation, if not more, than the pursuit of indefinite growth. Yet, as it has often been remarked, the vast consensus achieved around the term sustainable development depends on its intentional vagueness, not to speak of obfuscation.[49] For the unspoken subtext of sustainable development is of course unsustainable growth, that is to say the very problem the expression manages to mimetically cover up. Sustainable development is the wish fulfillment of a kind of growth that would be sustainable, even though the disavowal of the word itself suggests that growth is not compatible with sustainability. At worst, it is a contradiction of terms. Its success, like that of "green growth," relies on the denegation of the central conflict posed by an ecological economics. This conflict must on the contrary be at the heart of a reflection on limits.

Most dramatically, the recognition of limits forces us to confront the relation between growth and inequality. For many of the principled and thoughtful proponents of economic growth, often coming from the political left as well as from economic liberalism, the most compelling argument for continued expansion is that it is the indispensable condition for giving the "bottom third" of humanity a chance to rise from its still unrelieved material poverty. It is an argument whose moral strength is irrefutable, and if economic growth could be indefinitely sustainable, there would be no doubt as to the desirability of spreading its benefits to everyone. But once we accept the existence of limits, the argument turns around on itself. It is precisely because the poorer peoples of the world must have a chance to catch up with the rich that the latter should restrain their material consumption. And because we have already overshot the point of universal equilibrium by a long distance, this means that the most affluent among us must find a way to reduce our footprint so that others can increase theirs to a decent level viable for all. The rich have a human obligation, and not just an ecological one, to engage in *décroissance*.

It is of course a daunting task to even imagine how this could be achieved, but it is the only moral horizon on which the equality of human prospects can be thought in the era of finitude. What the perspective of limits makes us see is that growth has always been the alibi of inequality, perpetuated by the promise of a tide that lifts all boats.[50] Once universal growth is no longer possible, there is no place left to hide. We must either intentionally aim for an ecological footprint that will be equally achievable (and limited) for every inhabitant of the earth, or resign ourselves to a world in which the many are forever doomed to misery so that the few keep enjoying a greater share of the land. This manifesto, it will be noticed, does resemble and coincides in part with that of Marxism, though it is not reducible to a critique of capitalism per se, and stems from a conflict between humanity and nature that most Marxists, and the left in general, have long been loath to acknowledge, and have even resisted on theoretical grounds.[51]

To recognize that there is a conflict is of course not the same thing as knowing how it will end, or whether it can be resolved. However intellectually seductive we may find the model of historical dialectics, materialist or otherwise, conflicts do not always find a resolution, let alone one that we might wish for and find satisfying. The great forces that shape human history, especially those of economics and demography, operate to a large extent independently of human will and without, or in spite of, our consciousness. Rather than be overcome, this conflict may well overcome us. This is all the more possible as there are many other conflicts and emergencies competing for our attention, including famines and armed conflicts, or simply massive unemployment, that may always seem more urgent and immediately distracting, even if we were all persuaded that none is more vital for humanity. All powerful reasons for the many who are already convinced of the imminence of the crisis to have doubts about the efficacy of mere thoughts and words. And yet, even if it is true that all intellectuals, great and small, tend to overestimate the importance of ideas, there are also reasons to think that the failure of the world to act commensurately to the problem is still due in part to a deficit in intelligibility. For all the knowledge available to us, and all the immense intelligence and expertise brought to bear on different aspects of our predicament, we have difficulty making sense of it, let alone agreeing on what we must do.

Human beings need bread (or rice, or sorghum). They need recognition, and love. But they also need meaning, the only thing that can give us the conviction to embrace necessity. Not sufficient, but surely necessary if we hope to be prepared for the equally necessary messiness of politics and practical realities. This reflection is one attempt among many to contribute some meaning to the greatest challenge of our time, by raising the question of limits. It does not aim to say things that have not been advanced before, resting on the contrary on the scientific and theoretical work of countless others in the several fields that intersect in this question, including of course the environmental thinkers who are its main inspiration. Its most hoped for ambition would be to begin providing elements of a bridge, if not a synthesis, between those different disciplines, especially the competing ways of thinking and equally compelling arguments of ecology and economics, that so often seem to dwell on two separate planets. If it could be said unpretentiously, and without seeming to lay claim to yet another area of competence, this would aspire to be a philosophical reflection, not least in the sense that it will as often ask questions as provide answers. More justifiably, it should probably be approached as an essay in the history of ideas, since as we will soon see, the question of limits cannot be understood without going back to the period when it was first posed and challenged at the same time, the late Enlightenment that also gave birth to the modern idea of economic progress, and in particular to the seminal model of social evolution through stages of economic development, the "Four Stage Theory," that provided its original title. A historical reflection, then, as long as we remember that history is always also about the present.

Notes

1 Aristotle distinguished the management of the home, *oikonomia*, from which we improperly derive our name for economics, from the pursuit of monetary gain, which is *chrematics*. As Daniel Bell points out in *The Cultural Contradictions of Capitalism*:

> The art of household management [...] entails the observance of [...] natural limits. Acquisition has a limit in size, determined by the purposes (i.e. the natural needs) of the household. The acquisition that is unlimited [...] Aristotle calls *chrematistic*; it is "unnatural" precisely because it is unlimited.

And Bell adds a precious piece of etymological information: "In the literal Greek [...], the root *chremata* means things" (Bell, 1996, 223). See also note 37 (this chapter).

2 One the earliest and best-known figurations of this sense of limits is Kenneth Boulding's "The Economics of the Coming Spaceship Earth," first published in 1966 (Boulding, 1966). Referring to Boulding years later, Xenophon Zolotas expressed the sentiment becoming more common as:

> Mankind is becoming aware that [...] it has come close to the limits of earth's stock of energy and wealth. Beyond these limits, the effects of the inexorable laws of nature, such as the entropy law, will be highly intensified.
>
> (Zolotas, 1981, 134)

3 In addition to the appropriate scientific prudence in the face of uncertainty, James Hansen, one of the scientists most vocal about the dangers of climate change, deplores what he calls "scientific reticence" about going public among many of his fellow researchers. In Hansen's opinion, their understandable "concern about the danger of being accused of 'crying wolf' is more immediate than concern about the danger of 'fiddling while Rome burns'" (Hansen, 2009, 87).

4 Among the most vivid Pascalian evocations of this sense of infinity, we can cite:

> The whole visible world is but an imperceptible speck in the ample bosom of nature. No idea approaches it. We may swell our conceptions beyond all imaginable space, yet bring forth only atoms in comparison with the reality of things. It is an infinite sphere, the centre of which is every where, the circumference no where.

Or again:

> For after all what is man in nature? A nothing in regard to the infinite, a whole in regard to nothing, a mean between nothing and the whole; infinitely removed from understanding either extreme. The end of things and their beginnings are invincibly hidden from him in impenetrable secrecy, he is equally incapable of seeing the nothing whence he was taken, and the infinite in which he is engulfed.

And finally: "The eternal silence of these infinite spaces alarms me" [in the original, "m'effraie"], perhaps more strongly rendered as "frightens" or "terrifies me" (Pascal, 1901, 19, 28).

5 See Robert Henson: *The Thinking Person's Guide to Climate Change* (Henson, 2014, 56).

6 Joshua P. Howe gives a good example of the impact of this image, when an early MIT report on climate change, the "Study of Critical Environmental Problems" of 1970, was followed up a year later by another MIT report not only with the more arresting

title of "The Study of Man's Impact on Climate," but also with the iconic NASA photograph on the cover, with much greater effect (Howe, 2014, 57). We can similarly compare the various editions of the famous Meadows and Meadows essay *The Limits to Growth*, and *Limits to Growth, the 30 Year Update*, also graced with the same icon, with the more sober cover of the original Club of Rome report (Meadows *et al.*, 1972; Meadows *et al.*, 2004).

7 See Robert Socolow: "Climate Change and Destiny Studies: Creating Our Near and Far Future" (Socolow, 2015, 10–20).

8 This is related to *discounting*, the common economic practice of lowering by a certain yearly percentage the price of future goods. Similarly, it seems that we have a psychological propensity, no doubt personally and culturally variable, to attribute a declining value to the tangible or intangible goods we wish to have or fear to lose in the distant future. We suffer from what Avner Offer calls a "myopic vision of time" (Offer, 2006, Chapter 3).

9 *The Population Bomb* (Ehrlich, 1968). Malthus, the author of *An Essay on the Principle of Population* (first edition 1798), will be the subject of Chapter 3.

10 Paul Gilding gives as an example the list of physical limits identified by the Stockholm Resilience Center in *Planetary Boundaries*: "Climate change, stratospheric ozone, land use change, freshwater use, biological diversity, ocean acidification, nitrogen and phosphorus inputs into the biosphere and oceans, aerosol loading, chemical pollution" (Gilding, 2011, 227). Jared Diamond in *Collapse* proposes a list of 12 "most serious problems" (Diamond, 2011, Chapter 12).

11 See for example Karen L. Higgins: *Economic Growth and Sustainability* (Higgins, 2015, 84). 884 million people in the world, one in nine as of 2018, live without access to safe water (Water.org, 2018), and according to a detailed study of the world's 37 largest aquifer systems, between 8 and 11 were judged "overstressed" based on the ratio of "groundwater use to availability" (Richey *et al.*, 2015).

12 According to Thomas L. Friedman, "deforestation accounts for more than 20 percent of all CO_2 emissions." This is "greater than the emissions of the world's entire transportation sector – all the cars, trucks, planes, trains, and ships combined" (Friedman, 2008, 148).

13 According to the organization Global Fishing Watch, 90 percent of the world's fisheries are either overfished or fully fished (Global Fishing Watch, 2018). A World Bank report of 2013 projects that, as a consequence of growing demand, "aquaculture – or fish farming – will provide close to two thirds of global fish consumption by 2030 as catches from wild capture fisheries level off and demand from an emerging global middle class, especially in China, substantially increases" (World Bank, 2013).

14 "We are destroying the book of life before we have read it," writes Martin Rees: "The extinction rate is 1,000 times higher than it was before humans came on the scene, and it is increasing" (Rees, 2008, 42). According to the estimations of Conservation International, this averages to one species going extinct every 20 minutes (Friedman, 2008, 141).

15 Summarizing the findings of Glen P. Peters *et al.*, "Growth in Emission Transfers via International Trade from 1990 to 2008," Naomi Klein writes: "The researchers concluded that the rise in emissions from goods produced in developing countries but consumed in industrialized ones was six times greater than the emissions savings of industrialized countries" (Klein, 2014, 80). See also Adam Minter's *Junkyard Planet* (Minter, 2013).

16 Methane is 21 times more powerful than CO_2 as a global warming gas. In 2006, "[the] livestock industry accounted for 14–18 percent of total Greenhouse Gas emissions" (Higgins, 2015, 57).

17 For the difference between the ozone crisis and the efforts to reverse climate change, see in particular Joshua P. Howe (Howe, 2014, 150–154). For the rise and fall of CFCs, see Peter A. Victor: *Managing without Growth* (Victor, 2008, 75).

18 The latest Conference of Parties was held in Bonn, Germany, in 2017. The Kyoto Protocol was signed in 2005, and the Paris Accord in 2016.

19 *An Inconvenient Truth*, a documentary film directed by David Guggenheim about former US Vice-President Al Gore's campaign against global warming (Guggenheim, 2006). Al Gore and the International Panel on Climate Change were jointly awarded the Nobel Peace prize in 2007.

20 "Global Greenhouse Gas Emissions Data" (Environmental Protection Agency, 2018). According to Robert Henson, the Keeling Curve is "showing increasing CO_2 concentrations in the atmosphere and an exponential pattern of growth" (Henson, 2014, 39).

21 See the end of Chapter 5, as well as Chapter 6.

22 In addition to references in note 17 (this chapter), see Sharon L. Roan: *Ozone Crisis: The 15-Year Evolution of a Sudden Global Emergency* (Roan, 1989).

23 "A 2003 study by the ecologist Jeffrey S. Dukes estimated that the average gallon of gasoline requires over 90 tons of prehistoric biomass as raw material" (Hawken, 2017, 194).

24 Robert Henson notes that

> Not long ago, the concept of "peak oil" – the point at which we've extracted half of the world's accessible oil reserves – was gaining prominence … Lately, the concept of peak oil has come upon hard times. Estimates of accessible oil have jumped dramatically over the last few years with the advent of hydraulic fracturing (or fracking), as well as oil-sand extraction and other so-called unconventional ways of getting oil out of the ground.

He goes on to write, however, that "the concept behind peak oil remains valid" (Henson, 2014, 400).

25 Thus concern about peak oil, perhaps most intense in the early part of the twenty-first century, is gradually being replaced by the newer awareness that we cannot burn what is left in the ground for fear of the consequences for climate change. Ironically, the end of the oil age, like that of coal, would no longer be through exhaustion of stocks, but more similar to that of earlier energy sources which were replaced by more efficient ones. As Vaclav Smil points out: "transitions from wood to coal and from coal to oil were not brought by global exhaustion of the substituted resource, but by lower prices and better qualities of new fuels" (Smil, 2003, 195). Or in the oft-repeated *bon mot* attributed to Sheikh Ahmed Zaki Yamani at an OPEC meeting: "The stone age didn't end because we ran out of stones" (Friedman, 2008, 250).

26 Tim Jackson writes

> The third phase of the limits debate, is different from the last two. Resource Scarcity – the problem of "sources" in the language of environmental economists – is only part of the concern. The debate is driven even more strongly by the problem of "sinks" – the capacity of the planet to "assimilate" the environmental impacts of economic activity.
>
> (Jackson, 2009, 11)

"Even before we run out of oil," explains ecologist Bill McKibben, "we're running out of planet" (McKibben, 2007, 18).

27 For a comprehensive inventory of present and future technologies and their potential, see *Drawdown* (Hawken, 2017).

28 "[A] durable economy needs to be based exclusively on the regular influx of energy from the sun's radiation," wrote Hermann Scheer, one of the best-informed and compelling champions of decentralized renewable energy, pointing optimistically to the fact that the fraction of solar radiation that reaches the earth "is still 20,000 times more energy each day than mankind's current daily energy requirements" (Scheer, 2012, 17, 23).

29 [We] basically have two sources of low entropy: the solar and the terrestrial. They differ significantly in their patters of scarcity. The solar source is practically unlimited in its stock dimension, but is strictly limited in its flow rate of arrival on earth. The terrestrial source (minerals and fossil fuels) is strictly limited in its stock dimension, but can be used at a flow rate of our own choosing, within wide limits.

(Daly and Cobb, 1989, 196)

30 Paul Hawken and Amory Lovins are particularly enthusiastic, and sometimes persuasive advocates of such a technological transformation and its economic potential. See in particular *Natural Capitalism: Creating the Next Industrial Revolution* (Hawken *et al.*, 1999).

31 I am obviously using, without permission, the expression made famous by former US Vice President Al Gore as the title of the documentary film already mentioned. I hope it will be taken as a form of homage rather than theft.

32 The 1990s, according to Naomi Klein, "was also the period when Big Green became most enthusiastically pro-corporate, most committed to a low-friction model of social change in which everything had to be 'win-win'" (Klein, 2014, 210).

33 Brian Czech is particularly scathing and humorous on the abuse of the color green, when brown would be more appropriate: "We shouldn't be surprised if they start talking about 'green population growth' for green jobs and green consumerism." However, "[i]nstead of green growth, we have brown bloating" (Czech, 2013, 196–197).

34 Daniel Bell goes even further:

> The deepest nature of modern man, the secret of his soul as revealed by modern metaphysics, is that he seeks to reach beyond himself; knowing that negativity – death – is finite, he refuses to accept it. [... The] modern hubris is the refusal to accept limits.
>
> (Bell, 1996, 49–50)

35 The concept of disenchantment (*Entzauberung*), borrowed from Friedrich Schiller, was most famously used by Max Weber to describe the process of rationalization and instrumentalization of our relation to the world initiated by Western modernity, often associated with the rise of seventeenth-century science and the dualistic philosophy of René Descartes.

36 This does not mean, as we will have occasion to see in some detail, that early and classical economists were not to varying degrees conscious of limits, but that their concern was universally to expand or postpone those limits as far as possible. For a comprehensive and dispassionate examination, see W. W. Rostow: *Theorists of Economic Growth from David Hume to the Present* (Rostow, 1990).

37 See note 1 (this chapter). For Aristotle,

> the business of household management is concerned more with human beings than it is with inanimate property; it is concerned more with the good condition of human beings than the good condition of property (which is what we call wealth).
>
> (*Politics*, Book I, Chapter 13)

Daniel Bell comments:

> The basis of the [Aristotelian] *polis* is the satisfaction of natural needs. Unlimited acquisition can only be destructive of the household; the aim of household management, therefore, is restraint of desire. This is a point of view distinctly uncongenial to a democratic ethos and to the modern temper.
>
> (Bell, 1996, 252)

38 From the immense literature devoted to these questions, one would have to single out such different classics as Max Weber's *The Protestant Ethic and the Rise of Capitalism* and

Karl Polanyi's *The Great Transformation* (Weber, 1905; Polanyi, 2001), and of course the works of Marx and Engels and their many commentators.

39 See Jean Gadrey: "Le baromètre des inégalités dans le monde" (Gadrey, 2007), and for a more thorough analysis, Branko Milanovic: *Worlds Apart: Measuring International and Global Inequality* (Milanovic, 2005).

40 According to a prospective study by John Hawksworth and Danny Chan published by the professional services firm PwC, China "is expected to overtake the U.S. in purchasing power parity by 2030," and India by 2050.

> Given the rise of India and China, our model suggests that the US and the EU's share of world GDP in PPP terms will face a steady decline from around 33 percent in 2014 to only around 25 percent by 2050. This shift of global economic power to Asia may occur somewhat more quickly or slowly than this, of course, but the general direction of change and the historic nature of this shift are clear.
>
> In many ways, it is a return to the pre-Industrial Revolution era when China and India dominated world GDP in large part due to their great populations, and relatively efficient agricultural sectors at that time.
>
> (Hawksworth and Chan, 2015)

41 In his carefully documented article "Economic Growth," Max Roser produces a revealing graph of "World GDP over the last two millennia," based on World Bank data and estimates by the economic historian Angus Maddison. Strikingly, it is practically flat for almost the whole historical period, until about 1820, when the graph starts growing exponentially, the curve rising almost vertically: from $1.2 trillion in 1820 to $1.92 trillion in 1870, $3.42 trillion in 1900, $4.74 trillion in 1913, $7.81 trillion in 1940, $9.25 trillion in 1950, $14.62 trillion in 1960, $23.87 trillion in 1970, $34.73 trillion in 1980, $47.04 trillion in1990, $63.1 trillion in 2000, $91.33 trillion in 2010, and $108.12 trillion in 2015. A hundredfold increase in less than two centuries. The average per capita increase in GDP is of course not as high, given the explosion in world population during the same period, but it is still impressive, more than tenfold between 1820 and 2015, from 1,230 international dollars to 14,700 (Roser, 2018).

42 There are numerous criticisms of the inadequacy of the GDP as a measure of economic reality. For a concise summary, see Tim Jackson: *Prosperity without Growth* (Jackson, 2009, 125). For a full examination of the question of economic measurements: *Mismeasuring Our Lives: Why GDP Doesn't Add Up* (Stiglitz *et al.*, 2010).

43 Among the most important measures of economic welfare are: the Measure of Economic Welfare (MEW), developed by Nordhaus and Tobin in 1972; the Index of Economic Aspects of Welfare (EAW, Zolotas, 1981); the Index of Sustainable Economic Welfare (ISEW, Daly and Cobb, 1989); the Genuine Progress Indicator (1995), and the Happy Planet Index (HPI, New Economic Foundation). For an overview see Jean Gadrey and Florence Jany-Catrice: *The New Indicators of Well-Being and Development* (Gadrey *et al.*, 2006), and the European Commission website (European Commission, 2018).

44 Brian Czech refers in a strikingly similar formulation to "the self-sufficient services fallacy." Based on the very different biological model of trophic levels, his conclusions are substantially the same as those of the historical argument we will follow more fully in Chapter 5 on the basis of stadial and sectoral development. See *Supply Shock*, especially Chapter 7 (Czech, 2013).

45 The site Glassdoor indicates an annual base pay of $89,131 for financial traders. It does not provide a figure for "Additional Cash Compensation" (Glassdoor, 2018).

46 *More* is appropriately the title of a book by Robert M. Collins on the politics of growth in post Second World War America (Collins, 2000).

47 "Sublation" is one of the possible translations of the difficult but rich concept of *Aufhebung*, the term used by Hegel to describe the dialectical process (Hegel, 1807). In

Hegelian dialectics, an original opposition between two contradictory terms, say "growth" and "limits," is overcome, or resolved (sublated), by a third term that abolishes the contradiction, say "sustainable development." It is sometimes rendered as "sublimation," which would not be inappropriate for our example.

48 "Sustainable development is development that meets the needs of the present without compromising the ability of future generations to meet their own needs" (World Commission on Environment and Development, 1987, Chapter 2).

49 Joshua P. Howe offers a valuable historical perspective: "As economic historian Emma Rothschild notes, the concept of sustainability introduced in the so called Brundtland Report was a variation of a much older concept of "maintaining capita intact," an economic principle fleshed out by A. C. Pigou, Friedrich Hayel, and John Hicks in the 1940s, that cautioned against reducing capital stock for the sake of profit" (Howe, 2014, 172). "In the world of environmental politics," Douglas Torgerson comments in "The Uncertain Quest for Sustainability,"

> the ambiguity of sustainable development thus often seems attractive because it fosters enhanced co-operation among one-time opponents, the proverbial strange bedfellows of political life [...] The vagueness of "sustainable development," a weakness in terms of technical discourse, gains a certain political strength...
> (Fischer and Black, 1995, 11)

50 This will also be developed in Chapter 5. Under the concept of "dynamic egalitarianism," Fred Hirsch presents a particularly strong case for the same argument, and Richard Wilkinson and Kate Pickett summarize it perfectly in slightly different terms: "Growth is a substitute for equality" (Hirsch, 1976, 166–170; Wilkinson and Pickett, 2010, 226).

51 For an examination of Marx's thought on the question of ecological limits, see the first half of Chapter 3.

References

Bell, Daniel (1996) *The Cultural Contradictions of Capitalism*. New York: Basic Books.

Boulding, Kenneth (1966) "The Economics of the Coming Spaceship Earth," in *Environmental Quality in a Growing Economy*, H. Jarrett (ed.). Baltimore: Resources for the Future/Johns Hopkins University Press.

Collins, Robert M. (2000) *More: The Political Economy of Growth in Post-War America*. New York: Oxford University Press.

Czech, Brian (2013) *Supply Shock: Economic Growth at the Crossroads and the Steady State Solution*. Gabriola Island, DC, Canada: New Society Publishers.

Daly, Herman E. and John B. Cobb, Jr. (1989) *For the Common Good: Redirecting the Economy toward Community, the Environment, and a Sustainable Future*. Boston: Beacon Press.

Diamond, Jared (2011) *Collapse: How Societies Choose to Fail or Succeed*. New York: Penguin.

Ehrlich, Paul (1968) *The Population Bomb: Population Control, or Race to Oblivion*. New York: Buccaneer Books.

Environmental Protection Agency (2018) "Greenhouse Gas (GHG) Emissions" [online]. Available at www.epa.gov/ghgemissions.

European Commission (2018) "Beyond GDP: Measuring progress, true wealth and well-being" [online]. Available at www.beyond-gdp.eu.

Fischer, Frank and Michael Black (eds.) (1995) *Greening Environmental Policy: The Politics of a Sustainable Future*. New York: St. Martin Press.

Friedman, Thomas L. (2008) *Hot, Flat, and Crowded: Why We Need a Green Revolution – and How It Can Renew America.* New York: Farrar, Strauss & Giroux.
Gadrey, Jean (2007) "Le baromètre des inégalités dans le monde," *Alternatives Economiques*, no. 256, March.
Gadrey, Jean and Florence Jany-Catrice (2006) *The New Indicators of Well-Being and Development.* Basingstoke: Palgrave Macmillan.
Gilding, Paul (2011) *The Great Disruption: Why the Climate Crisis Will Bring on the End of Shopping and the Birth of a New World.* New York: Bloomsbury Press.
Glassdoor (2018) "Financial trader salaries" [online]. Available at www.glassdoor.com/Salaries/financial-trader-salary-SRCH_KO0,16.htm.
Global Fishing Watch (2018) [online]. Available at http://globalfishingwatch.org/.
Guggenheim, David (2006) *An Inconvenient Truth.* Documentary film, Paramount.
Hansen, James (2009) *Storms of my Grandchildren: The Truth about the Coming Climate Catastrophe and our Last Chance to Save Humanity.* New York: Bloomsbury USA.
Hawken, Paul (ed.) (2017) *Drawdown: The Most Comprehensive Plan Ever Proposed to Reverse Global Warming.* New York: Penguin Books.
Hawken, Paul, Amory Lovins and L. Hunter Lovins (1999) *Natural Capitalism: Creating the Next Industrial Revolution.* Boston, New York and London: Little, Brown and Company.
Hawksworth, John and Danny Chan (2015) "The World in 2050: Will the shift in global economic power continue?" [online]. Available at www.pwc.com/gx/en/issues/the-economy/assets/world-in-2050-february-2015.pdf.
Hegel, G. W. F. (1807) *The Phenomenology of Spirit.* Oxford, New York, Toronto and Melbourne: Oxford University Press (this edition 1977).
Henson, Robert (2014) *The Thinking Person's Guide to Climate Change.* Boston: American Meteorological Society.
Higgins, Karen L. (2015) *Economic Growth and Sustainability: Systems Thinking for a Complex World.* Academic Press, Elsevier.
Hirsch, Fred (1976) *Social Limits to Growth.* Cambridge, MA: Harvard University Press.
Howe, Joshua P. (2014) *Behind the Curve: Science and the Politics of Global Warming.* Seattle and London: University of Washington Press.
Jackson, Tim (2009) *Prosperity without Growth.* London and New York: Earthscan-Routledge.
Klein, Naomi (2014) *This Changes Everything: Capitalism vs. The Climate.* New York: Simon & Schuster.
Malthus, Thomas Robert (1798) *An Essay on the Principle of Population.* Cambridge, MA: Cambridge University Press (this edition 1998).
McKibben, Bill (2007) *Deep Economy: The Wealth of Communities and the Durable Future.* New York: Holt.
Meadows, Donella H., Dennis L. Meadows, Jørgen Randers and William W. Behrens (1972) *The Limits to Growth: A Report for the Club of Rome's Project on the Predicament of Mankind.* New York: Universe Books.
Meadows, Donella H., Jorgen Randers and Dennis Meadows (2004) *Limits to Growth: The 30-Year Update.* White River Junction, VT: Chelsea Green Publishing Company.
Milanovic, Branko (2005) *Worlds Apart: Measuring International and Global Inequality.* Princeton, NJ: Princeton University Press.
Minter, Adam (2013) *Junkyard Planet: Travels in the Billion-Dollar Trash Trade.* New York, London, New Delhi and Sydney: Bloomsbury Press.
Offer, Avner (2006) *The Challenge of Affluence: Self-Control and Well-Being in the United States and Britain since 1950.* Oxford: Oxford University Press.

Pascal, Blaise (1901) *The Thoughts of Blaise Pascal*, translated from the text by M. Auguste Molinier by C. Kegan Paul. London: George Bell and Sons.

Polanyi, Karl (2001) *The Great Transformation: The Political and Economic Origins of Our Time*. Boston: Beacon Press.

Rees, Martin (2008) "Science: The Coming Century," *The New York Review of Books*, November 20.

Richey, Alexandra S., Brian F. Thomas, Min-Hui Lo, John T. Reager, James S. Famiglietti, Katalyn Voss, Sean Swenson and Matthew Rodell (2015) "Quantifying Renewable Groundwater Stress with GRACE," *Water Resources Research*, vol. 51, no. 7, 5217–5238.

Roan, Sharon R. (1989) *Ozone Crisis: The 15-Year Evolution of a Sudden Global Emergency*. New York, Chichester, Brisbane, Toronto and Singapore: Wiley Science Editions.

Roser, Max (2018) "Economic Growth" [online]. Available at OurWorldInData.org, https://ourworldindata.org/economic-growth.

Rostow, W. W. (1990) *Theorists of Economic Growth from David Hume to the Present: With a Perspective on the Next Century*. New York and Oxford: Oxford University Press.

Scheer, Hermann (2012) *The Energy Imperative: 100 Per Cent Renewable Now*. Oxford and New York: Earthscan.

Smil, Vaclav (2003) *Energy at the Crossroads: Global Perspectives and Uncertainties*. Cambridge, MA: MIT Press.

Socolow, Robert (2015) "Climate Change and Destiny Studies: Creating Our Near and Far Futures," *Bulletin of the Atomic Scientists*, vol. 71, no. 6, 18–28.

Stiglitz, Joseph, Amartya Sen and Jean-Paul Fitoussi (2010) *Mismeasuring Our Lives: Why GDP Doesn't Add Up*. New York: The New Press.

Victor, Peter A. (2008) *Managing without Growth: Slower by Design, Not Disaster*. Cheltenham, UK and Northampton, MA: Edward Elgar.

Water.org (2018) "The Water Crisis" [online]. Available at https://water.org/our-impact/water-crisis/.

Weber, Max (1905) *The Protestant Ethic and the Spirit of Capitalism*. Oxford: Routledge (this edition 2001).

Wilkinson, Richard and Kate Pickett (2010) *The Spirit Level: Why Greater Equality Makes Societies Stronger*. New York: Bloomsbury Press.

World Bank (2013) "Fish to 2030: Prospects for fisheries and aquaculture (English)." Agriculture and environmental services discussion paper; no. 3. Washington DC: World Bank Group [online]. Available at http://documents.worldbank.org/curated/en/458631468152376668/Fish-to-2030-prospects-for-fisheries-and-aquaculture.

World Commission on Environment and Development (Brundtland Commission) (1987) *Our Common Future*. New York: Oxford University Press.

Zolotas, Xenophon (1981) *Economic Growth and Declining Social Welfare*. New York and London: New York University Press.

2 The last stage

Scarcely more than 200 years ago, a turning point was reached in the history of limits. An event so recent in the span of human experience, let alone of life on earth, that it may one day appear as merely a glitch in our relation to the world we live in, to our being in the world. Yet it is a turn that has since then proved so compelling, whose effects have become so pervasive, that we are already hardly able to think, after such a short period of time, outside the horizon it opened up for humanity.

There is of course nothing exceptional in locating around the turn of the eighteenth century a major shift in the history of the West, and indeed of the whole world that would increasingly come under its influence and domination. The decades surrounding the year 1800 have long been recognized as a time of momentous upheaval in the aesthetic, political, and socio-economic conceptions of the Eurocentric world, coinciding as they do with such variously interrelated movements as the birth of romanticism, the political revolutions in Europe and the Americas, and the beginnings of the industrial revolution itself. Philosophically, it straddles the fault-line between the Kantian culmination of Enlightenment rationality and the Hegelian elevation of History as the henceforth fundamental mode of Being, and it is no surprise that when Michel Foucault, much closer to us, sought to signpost an archaeology of Western knowledge, he too situated at that point the least controversial of his epistemological breaks.[1]

The event in question is undoubtedly connected with several of those widely recognized shifts, and in some sense subsumed, if not directly caused by them. But at the same time as being merely one aspect of larger socio-economic and ideological transformations, it can also be understood, in a more complex causal relation, as the underlying ground of possibility of this new outlook of modernity, even if it remains largely invisible to us by virtue of its very obviousness. This turning point can be thought as the breaking, or at the very least the indefinite postponement, of the barrier of limits. For the first time in the history of Western civilization, and we can venture to say with some assurance, in the secular experience of all other civilizations past and present in the world, human beings dared to think not only of the desirability, but also of the actual possibility of a progress without limits.

More particularly, since this is the specific form of progress that concerns us here above all else, the idea began to spread that the wealth of nations, the ability of human societies to increase their level of production and consumption of earthly goods, could continue to expand indefinitely, to keep growing without assignable limits. Indeed, it is from this time that what we have come to know as the economy has become inseparable from the normative notion of infinite growth. A good economy will henceforth be a growing economy, not only in its infancy and youth, but also in the healthy state of its never-ending maturity, thus escaping from the traditional analogy with mortal life-cycles. Economic progress has come to be understood as a naturally limitless process, a progression without end. In a leap of almost unimaginable boldness, modernity has theorized, here on earth, the practical transcendence of finitude.

No doubt, many cultures and civilizations before Western modernity had dreamt of a prosperity without end, of infinite plenty for their descendants to come, in this world or another. Historically, the masters and chroniclers of the great empires that in different periods and places came to rule the whole worlds known to them, and for much longer times than our recent globalizing modernity has so far lasted, could surely be tempted to believe that their expansion and power would know no limits, and were even destined to last for eternity. Metaphysically, and this is a point of much greater importance that we will have to come back to when we reflect on the question of goods, several great spiritual traditions, including the ones Western modernity is heir to, have long associated limitlessness with the idea of perfection, and placed infinity among the highest attributes of the divine. It may perhaps even be argued that theologically, long before the development of a separate philosophical tradition, and contrary to common sense expectations, the notion of limitlessness preceded that of limitedness, the former precipitating the latter in the consciousness of human finitude.[2] There is at any rate hardly a doubt that for most cultures and people – though there are very important exceptions to this – to be without limits is considered a higher good than to be confined within limits, or conversely, that to be limited, as opposed to being unlimited, is a flaw, a negative condition. To be limited, in short, is considered a limitation, and the desire to transcend this limitation has long been one of the most valued human aspirations, at least in the spiritual realm.

Still, even while recognizing the full weight of the above considerations, it is imperative to remind ourselves that here on earth, most human beings for almost the entire history of our species have known how to live within limits, and thought of their existence as being ruled by natural thresholds that cannot be crossed without peril. All traditional agricultural peoples – and Europeans until the end of the eighteenth century were still largely agricultural in this sense – know about the unchanging if unpredictable rhythms of seasons and the confines of land and plant productivity, and cannot afford to ignore those limits, no matter how hard they might try. This was even more true during the hundreds of thousands of years when our ancestors lived by collecting the plants spontaneously produced by the earth on its own time and killing the animals it naturally sustained.[3] The apparent nonchalance that has often been observed among hunting and gathering peoples about ensuring their livelihood, the sense of security and affluence that so astonished Western travelers frequently harried in the same environment, is precisely due to an exquisite knowledge the former had developed about the productive limits of their world, and their care to stay well below natural thresholds of sustainability. They seemed all the more improvident because they knew to leave a comfortable safety cushion between their needs for subsistence and the limits of natural replenishment, deliberately restraining from pushing those to the limit.

What the development of agriculture means – the development of traditional agriculture, from its gradual emergence some 10,000 years ago, Jean-Jacques Rousseau's aptly named "first revolution,"[4] to be distinguished from the much less aptly named "green revolution" of the twentieth century, which is really the industrialization, or what we should also call the fossilization of agriculture – what this first revolution consists in, the intentional harnessing and control of the organic processes occurring spontaneously in nature in order to amplify and accelerate them, amounts to pressing on those limits.[5] Unlike gathering and hunting, agriculture artificially pushes natural production to the limit. Hence its miraculous power to multiply the means of sustenance, and to support substantially larger and more concentrated human populations, but also (contrary to common perception) its much greater precariousness, should those limits be crossed. The ancient peoples who first experienced this miracle understandably worshipped the fertility of the earth, together with the radiance of the sun and the flooding of the waters that made this pregnancy possible, as well as the magnificent tamed beasts that provided the power to move and plow, and the manuring of the fields. Even as the first great city-states and empires proudly grew to sizes and densities that only intensive irrigated agriculture could sustain, they remained in awe of those primeval forces on whose mercy or wrath they were wholly dependent.

Gilgamesh, the legendary founder of one of those proud cities, still provides moving testimony of the conflict between the hubris of the first conquering civilizations and their humbling sense of natural limits. The epic is now primarily read as the failed attempt of the hero to transcend one of those limits: his quest for eternal life.[6] But it also chronicles the taming of wildness on which civilization depends, and most strikingly, the fateful crossing of another limit, in which we can with hindsight recognize the cause of its future downfall as well as the source of its greatness: the conquest of the cedar forest. The hero himself hesitates before accomplishing the murderous deed, as if aware of transgressing the limit between sustenance and destruction. We now know that the unassailable walls of Uruk, on which Gilgamesh sought to bestow his impossible eternity, have all crumbled into dust, as so many other great cities in the once Fertile Crescent. They fell from the blows of other rising city-states, that in turn fell as other empires rose. But more fundamentally, they fell from the neglect of the greater powers from which they had grown: from the sterilizing salinization of the land, from the drying out of the waters, from the desertification of the once endless forest. They fell – as several other early agricultural cultures across the world also would during the following millennia – for having ignored what no agricultural society can afford to cross: the limits of the natural sustainability of the earth.[7]

In the eighteenth century, the cuneiform tablets on which the story was written still lay buried in the rubble of those ancient clay cities. Europeans did not know yet about this particular civilization, nor about the full causes of its demise. But they did know about ruins, and about the rise and fall of other ancient civilizations, and of the waning and destruction of even the greatest empire the world had ever known, the Roman civilization in which they recognized their own cultural origins, and to which they still compared themselves with humility, even as they imagined they were bringing it partly back to life, still overwhelmingly under the sway of its aesthetic, literary, and political models. We justifiably see the Enlightenment as future oriented, in recognition of all the unprecedented knowledge, principles, and altogether new concepts it contributed to the world, one of which is the subject of the present reflection. And there is no doubt that many of its representatives since the seventeenth century were perfectly conscious that they were breaking new ground, and proudly so. But this should not make us forget the weight they felt from the past. Even for the resolutely progressive thinkers and scholars contributing to that great monument to modernity which is Diderot's *Encyclopedia*, it had been a mere 200 years since the rebirth of "the sciences and the arts," a fragile span of time on the scale of a millennial history still haunted by the trauma of the dark ages. The Enlightenment still knew that civilizations are mortal. One of its other great late summations is famously devoted to the *Decline and Fall* of that late empire it was still unsure to match either in grandeur or in extension, let alone in longevity.[8]

As for its view of the two contemporary civilizations Europe could compare itself to, the Ottoman Empire and China, as living examples of regression and stagnation respectively, it is of course difficult for us not to see it as hopelessly tainted by prejudice, and by the deplorable ethnocentrism that would soon assume blatant form in the Hegelian philosophy of history and in nineteenth-century racialism. It is not certain, though, that this view expressed Europe's arrogance and sense of superiority as much as its pessimism as to the fate of all civilizations, and the abiding hold of a cyclical conception of history it inherited from Antiquity and still shared with many other cultures.[9] As to belief in progress, if we hope to catch the dominant mood at the mid-eighteenth century, just on the eve of the major turn the West was about to take, there is probably no better guide than the sober, humane, and evenly balanced skepticism of the wise Montesquieu, one of the most universally respected figures of the age.[10] To everything, there is a season, and what goes up, here on earth, will most probably come down.

And for a sense of the limits imposed by nature on human beings living from the land, we would also do well to revisit his much misunderstood theory of climates, which is really a reflection on the primacy of place, on the particularity of geography, of what we now call the environment. Only modern people who have forgotten to pay attention to where and what they live on (or another great skeptic like Hume, who was already arguing for a world without limits), would find it odd to think that the place and the land and the climate we inhabit influence what we do, and who we are. Eighteenth-century Europeans in their overwhelming majority, as indeed almost all human beings living then in the world, still knew about the constraints of the land, and the limits of natural productivity. On average about three quarters of the population were still directly engaged in agricultural work.[11] Among those privileged enough to live from the profits of the land, and those (sometimes the same) who took an interest in the welfare of nations, most notably the fairly new breed of philosophers who called themselves Oeconomists, and particularly the French variety known as the Physiocrats,[12] many were seriously involved in agrarian reform and the scientific pursuit of agricultural improvement. They did not need to be reminded about limits. Limits were what they were up against in their efforts to produce more food to feed more people, still considered the primary measure of the wealth of nations. The means at their disposal, as from the beginning, were no more than the natural, if selected fertility of plants, the present-time energy of the sun, almost solely the power provided by the animals nourished by the same land, fertilized only by the recycled plant by-products and waste from the same animals bred for labor or food. The limited means of renewable flows.

It has often been remarked that the most materially advanced societies until the eighteenth century were fundamentally based on the same resources and technologies as those of classical Greece, or for that matter, ancient Mesopotamia. That is not quite true. Among the factors relevant to agricultural production and use, they benefited from significantly improved plowing and harnessing technologies, as well as the windmill unknown to Antiquity and waterpower technologies that qualify as pre-industrial,[13] and for transportation and distribution, much better seagoing ships and navigational techniques, as is well known from the already global scope of early modern economies. But the fact remains that for all the immense progress in pure science and theoretical thinking that had been made since the seventeenth century, and the explosive potential of the technological applications that were being developed at an unprecedented rate, late eighteenth-century societies still ran almost entirely on the real-time energy of the sun, or from the seasonal productive cycles of the earth and the wind and water power that all depend directly on solar power (the only ultimate source of energy we earthlings have, with the exception of the mini-sun that constitutes the heart of our own planet). In other words, since in retrospect this decisive difference can only be encapsulated for us in the negative, nobody anywhere had yet started using any significant amount of fossil energy.[14] People did use minerals of course, including metals, whose formation dates back hundreds of millions of years – that uniquely human entropic process had started as early as the paleolithic. But as long as the only sources of energy available to extract and transform and transport such materials are animal, water and wind power, the pace and scope of possible activity are necessarily relatively modest. The only very large-scale transformation of the planet associated with traditional agriculture and its primarily solar energy base that is unidirectional and partly irreversible is the destruction of forests for plantation, timber, and fuel: Gilgamesh's deed, the first mythical crime against nature.

In such a world, on an already long settled and developed continent, the sense of limits is ever present and inescapable. Progress and growth are slow and incremental, and are always at the mercy of a setback. In Europe, overall population, that most valued measure of prosperity, had rebounded and reached new heights after the dramatic collapse of the fourteenth century, but it had remained fairly stable since the seventeenth, and was only beginning to pick up again in the second half of the eighteenth century. It is telling, for an appreciation of perceptions if not of realities, that for several decades following that mid-century point a majority of observers and concerned policy-makers, trusting the authority of an opinion expressed by Montesquieu, were persuaded that European populations were *declining* when they were in fact growing, and it took a long time and much controversy for that impression to be recognized as erroneous.[15] Even then, on the eve of the greatest demographic surge the world had ever known, it seemed more natural to believe in the limits of human expansion that in its reality.

And last, before turning to the radically new paradigm that was being opened, largely by economists, at the very same time, it is also helpful to remember, even if at the risk of simplification, what the previous dominant school of economic thought had been concerned about. For the Mercantilists, as they have come to be known, the wealth of nations was represented by the finite amount of precious metals that could be extracted from the earth, and international competition for that wealth was conceived of as a zero-sum game. What the winners gained, the others lost. Even the realm of economics, the imagination of wealth itself, was constrained by the notion of limits.

It is against this background, literally against its constraints, that we must situate the emergence of the model of unlimited growth which we have in turn inherited, and whose consequences we are only now beginning to confront. Only then can we fully appreciate its radical audacity and its immensely liberating appeal at the time when it first appeared, but also its historical and cultural determinacy, and, in the long scheme of human history, how recent this model really is, and perhaps how ephemeral its sustainability.

Although a conceptual shift of such importance cannot be attributed solely to any particular individual, as a whole complex web of contributing factors and trains of thought undoubtedly led to its formulation, and the very suddenness and near unanimity of its acceptance, within a few decades, cannot help make us feel that it was in some sense inevitable, it is nevertheless widely recognized that it found its earliest fully developed expression in the pioneering work of two particularly prescient thinkers, both justly famous at the time, if no longer equally so nowadays, Adam Smith in Britain, and Turgot in France.[16] It would therefore not be wrong to identify this event with their names, as long as we see them as preeminent representatives and exponents of the new framework, among many others, rather than its unique authors, or in Foucault's striking phrase, as founders of discursivity.[17] While both Smith and Turgot certainly had distinctive and different personalities and inclinations, which led them to emphasize particular notions in their individual work, it is fair to say that these particularities did not leave a permanent mark on the future history and acceptance of the concepts they propounded, in the way that they would with the next major figure we shall have to encounter in the history of limits, that of Malthus. If anything, among the other important precursors of liberal economics in the mid-eighteenth century, the powerful personalities of Dr. Quesnay, the father of the Physiocratic movement in France, or of David Hume, himself a mentor to Adam Smith, may have had a more idiosyncratic impact. But significant and interesting as these individual variations may be, they are less important for the future of the idea than the very large agreement between them.

What they jointly and famously affirmed (in this, there were not alone, nor even the first) is as unremarkable to us, so familiar has it become to modern habits of thought, as it was provocative to traditional mentalities: that in a modern economy, it is enough to give free and unrestrained rein to the exercise of private interest for society to increasingly achieve the greatest prosperity. In the regime of perfect liberty – the original formulation of what we have come to know as economic liberalism – the unrestricted play of all individual inclinations will naturally combine to result in the optimum growth in productive activity and exchange. All that governments have to do, in the famous phrase, is to "laisser faire"[18] private initiative, and "laisser passer" all goods without duties or restraints, and the natural incitement of human needs and desires will necessarily lead to ever greater wealth, as if guided, suggested Smith equally famously, by a providential, if invisible hand. For it is in the very nature and logic of human desire to be infinite. We always want more than we have, because we can only want what we do not have. An economy driven by private interest, and giving free range to its full potentiality, will therefore be without limits.

Is that all they were saying? And would such liberalization of individualism have been sufficient for any society to achieve limitless prosperity? Of course not. But before we look at the economic conditions, and perhaps even more so, at the intellectual horizon under which such thinking could become possible, let us acknowledge what an extraordinary shift the rehabilitation of self-interest represents in the secular experience of Western culture. The privileging of individual desire does nothing less than turn on its head the whole of classical morality. Indeed, it may even relegate it altogether to the department of Antiquities. What for the whole classical tradition, as far as we may want to see it go back into the past, had represented the greatest threat to social cohesiveness, as well as the primary source of self-delusion and unbalance, now became encouraged for its social benefits, and lauded as a legitimate personal motivation. The problem of what were called the passions, perhaps the most central preoccupation of moral reflection throughout the age of reason, and the subject of some of its greatest art and literature, has not only lost its centrality, but ended up providing the solution.[19]

Not that either Smith or Turgot are ever tempted to make a virtue of selfishness. That was the provocative gesture of Mandeville's cynical *Fable of the Bees*, for which Smith has nothing but contempt. And as a professed admirer of La Rochefoucauld (Turgot, with a French classical education and an early training for the priesthood would have been equally steeped in this moralist tradition) Smith was well aware of the profound deceitfulness of *amour propre*.[20] In his own *Theory of Moral Sentiments*, his exposition of the delusions and vanity of selfish human endeavor is worthy of the most somber seventeenth century meditations on the subject, if not positively Jansenist.[21] But then he dramatically turns around to conclude: "It is well that nature imposes upon us in this manner" (Smith, 1759, 183), for to this vanity we owe most of human activity and progress and the beneficial pursuit of wealth. The providential hand is not only invisible, it is also cunning. It uses the lure of our imagined enjoyment of fortune, and our pride in the possession of "baubles and trinkets," which every wise person knows will never make us happier, for the greater prosperity of society as a whole. Selfishness may not be a moral good in itself, but it serves the more objective and larger good of economic progress.

From this derives a new version of the "double truth" at the heart of modernity. In the private care of ourselves and our loved ones, in seeking solace for the sorrows and inevitable losses of life, as well as in our altruistic concern for humanity itself, we may well choose to practice the old art of moderating our passions, of weighing the gratification of a present satisfaction against the benefits of a future one. We may even enjoy the bittersweet pleasures of postponement; even, who knows, the possible rewards of renunciation. If we are fortunate enough, we might end up relearning what thoughtful human beings in all cultures and walks of life have always known from time immemorial, that the better part of desire is often in the desiring itself, rather than its fulfillment. But as economic agents, or perhaps more properly as subjects to the economy, we are only supposed, indeed encouraged, to fulfill our desires as much as we can. The better to consume, we had better be consumed by desire, or better yet, take our desires for actual needs. Hence, logically, the hyper-development we are very familiar with in our advanced economies of an incessant incitement to want more, our constant interpellation, *qua* economic subjects, as desiring machines.

There are reasons to think that Smith himself would have been appalled by this development, he who favored at least a minimum of universal education to counterbalance the power of merchants and merchandizing. Smith was even aware of a deeper contradiction, since the virtue most valued for capital accumulation, the key to economic expansion, is *restraint*, the delayed gratification necessary for saving, rather than expense. But that may be yet another double truth inherent in the psychology of modern economies. Whatever the reservations Smith may have had about the system he was advocating, and the personal dislike he would probably have taken to its later evolution, there is no doubt that he contributed as much as his fellow travelers to the rehabilitation of self-interest, and its enthronement as the primary motivation and drive of economic progress, with two perhaps unforeseen and even unwanted consequences that have turned out to be crucial for the construction of modernity. One is the reduction of *homo economicus* to a self-interested individual, *as if* when we act as economic agents, we were only motivated by our self-interest, a self-fulfilling and as it were performative fiction that is increasingly leading people to think that human beings really are mostly self-interested individuals. An impoverishment of the conception of human subjectivity that most pre-modern cultures would have regarded as pitiful. The second consequence, in a sense the reserve side of the first, is that modernity, as a project committed to the notion of the self-interested individual (as possibly distinct from the actual persons who live in it) has turned its back on the wisdom traditions of the world, i.e. on the reflection on desire and on the respective merit of the goods to be pursued in this life, for others as well as for ourselves. Which is pretty close to saying that it has renounced morality.

These are very serious consequences, and for all of us who are within modernity, embracing many of its values and benefits, and who in any case cannot think that there is any outside of modernity except in utopian or fundamentalist rejectionism, they are grave concerns that we must eventually find a way to address. But how and why did the founding economists and philosophers of the late Enlightenment, who were certainly no less thoughtful nor moral than anyone before them, or than most of us, come to believe that the daring choice, indeed the huge gamble of liberating self-interest from all its traditional strictures was not only possible, not only worth it, but that its time had come? What did they see in their own historical and cultural moment that warranted such a dramatic shift? The short answer, as we know, is what we now call capitalism, or free market economy, or liberalism. And we might be content with that answer, since it is substantially right, except that once again its retrospective obviousness, its present evidence to us may make us lose the sense of its power and historical significance. It still does not tell us why the advent of the new economy seemed so imperative to them, indeed why it imposed itself with such finality. In order to retrieve this sense, we must again become conscious of the still larger intellectual framework within which capitalism came to be thought in the eighteenth century, a model we have also inherited from its great early theorists, sometimes without being fully aware of it.

For the founding fathers of liberal economic theory were not just economists in our modern sense, nor quite philosophers in the specialized meaning the term has acquired today, but rather in the very broad way the French used it at the time. They were intellectuals and theorists at large, who made important contributions in a wide range of fields that were then much more continuous and less disciplined than now. And in the case of Turgot and Smith, much as they deserved their future fame as economic forefathers, their foundational legacy may lie equally in what we now call anthropology, or cultural history. There is at any rate no doubt that their strictly economic thought (for us) was for them inseparable from a sweeping view of human evolution, and an epoch-making theory of universal history they were largely instrumental in elaborating and promoting.

This anthropological model of cultural progress, whose first mature formulation was arrived at almost simultaneously by Smith and Turgot in the early 1750s (Rousseau is another candidate for early formulation, but his version is as usual much more idiosyncratic), has been aptly described by Australian/British economic historian Ronald Meek, to whom we owe its compelling genealogy in *Social Science and the Ignoble Savage*, as the "Four Stage Theory" (Meek, 1976). It is a theoretical model that has since then proved so triumphant, and provided the paradigm for so many future schemes of progress, that it is familiar to all of us. In its generic form, the Four Stage Theory states that all human societies necessarily pass (or have already passed at some pre-historical time or will eventually pass at some future time) through the same successive stages of material development, in the same order, namely, in Smith's terminology: first "the Age of Hunters" (what we usually call hunting and gathering), second "the Age of Shepherds" (pastoralism), third "Husbandry" (agriculture), and finally fourth, the stage of what both Smith and Turgot call "Commerce"[22] (we would now identify it as Industrial, a decisive difference we will come back to in a moment).

In a long history of intellectual models, of what Michel Foucault (himself the author of one of the last great stage theories) called *epistemes*, it would be difficult to overemphasize the importance of this reconceptualization, which truly qualifies as a paradigm shift. The diversity and incommensurability of human societies, which had since Antiquity always been apprehended in the spatial dimension of a geography, and ordered according to their distance from a presumed center of civilization – the rearrangement of the human world known to the West in the Renaissance in a new continental distribution is only a confirmation of this enduring privilege of space – will from now on be classified and ordered in the single temporal dimension of a universal history, each human society (past or present) fitting at one of the successive stages. A shift from space to time, and from geographical location to stadial progression, that would become the veritable matrix of all master narratives and evolutionary theories to come throughout the nineteenth century, no matter the number and defining characteristic of their chosen stages, and inaugurate the reign of historicity as the fundamental mode of being of modernity.

Retrospectively at least, it is hard to imagine that the idea of limitless expansion could have been thought without this as it were anterior shift (though they are in their actual genesis simultaneous and inseparable) from the spatial to the temporal dimension, since the openness of time would seem to be the necessary ground of possibility for breaking the barrier of limits (always understood as a limitation of physical space). As we recalled with some insistence at the beginning, so long as one kept thinking in the spatial terms of the agricultural field, of the extension and carrying capacity of arable land, even of all possibly cultivated area in the world, one kept coming up against the inescapability of limits.

A truth the next generation of economists were soon to rediscover with the Ricardian law of rent and diminishing returns.[23] Although in the second half of the eighteenth century the geographical world still looked to European eyes almost infinitely open to exploration and (colonial) expansion, we know that the process of discovery and conquest that had started 300 years before was in fact about to reach its final frontiers, and that the globe was about to close in on itself. Until the tiniest of cracks would be opened again in the infinity of space in the late twentieth century (but the sky is still our limit), only in the direction of time could the possibility of limitlessness, and infinite possibilities, be thought. Which is why, to give a peripheral example, the tradition of classical utopias that had dominated the European political imagination since the Renaissance – almost always located on some undiscovered island – gradually gave way to that of uchronias, ideal cities projected in the distant future, or why, less peripherally, the nineteenth century saw a massive return of eschatology, in the form of secularized grand narratives of human progress. In spite of their equivalence for theoretical physics, in the common imaginary of modernity, space is limited; only time remains open to infinity. There is in any case hardly a doubt that the idea of limitless economic growth could only be thought in the open-endedness of time, and it is no coincidence that the thinkers who first dared to imagine such a possibility were also the first to formulate with clarity and assurance a completely temporal vision of human history as a universal process of unending progress.

But progressive time is not the only salient feature of the Four Stage Theory. The defining characteristic of a stage theory of progress is to be *discontinuous* as well as progressive.[24] Each stage, identified with a distinct mode of production (hunting and gathering, pastoralism, agriculture, etc.), is conceived of as a state, that is to say as more or less static during a certain period of time, sometimes during a very long period. It is in discontinuity, in the passage from one stage to the next, that progress occurs. The whole question of progress in a stage theory is to explain how and why society moves from one stage to another, what motivates, or necessitates, or prevents such a move. In short, to explain why each state, however long it might have lasted in actual historical time, is only a temporary stage in a universal progression. And such was indeed the great debate, and the great divide, unequal as it turned out to be, among the early exponents of the theory. For Rousseau, that most reluctant of progressivists, each successive stage was so stable, so self-sufficient (even if decreasingly so), that humanity ought never to have left it, so that it was impossible to imagine how each passage could ever have happened except through extraneous, accidental causes, as a fateful revolution conceived of in the mode of catastrophe. For Smith and Turgot, on the contrary, as for the great majority of their followers, the early stages are so inherently unstable, so precarious in their deficiency (even if decreasingly so), that they could hardly have existed at all, and what is difficult to imagine is how they could, as in fact historically they did, have lasted so long.

That is, so far as the early stages are concerned. For there is, in the four-stage theory as it is formulated in the later eighteenth century, another kind of discontinuity, more subtle and perhaps less immediately apparent, but no less crucial for the question of limits. The first three stages – this is a decisive departure from earlier theologically based antecedents of the theory, or that of a predecessor like Vico[25] – are clearly stages of material production, or subsistence, defined by the way in which people primarily make their living: hunting and gathering, pastoralism, agriculture. But the fourth, and last stage, "commerce," is not exactly a mode of *production* in the same sense as the others, and as our later conceived fourth stage of industrialism will also be, nor is it some vague adumbration of that coming technological revolution, much as the two developments may now appear to be related.[26] What is being recognized and promoted to the culminating stadial position for its unique capacity is a mode of exchange and circulation, but even more of reproduction and proliferation of wealth. Commerce, of course, with a new emphasis on the benefits of free and open trade, the market side of our market economy, but more decisively – what Antiquity and the pre-modern world had never been able or willing to recognize and take full advantage of – the miraculous power of wealth to breed ever more of itself through the logic of capital investment and return on investment producing more capital for reinvestment. That is, the capital side of our capitalist economy, though it did not yet call itself so. What is unique about this mode, in contradistinction to that of the previous stages, is not only its greater immateriality, but also and especially (though these two aspects are not unrelated), its potential infinity. In that sense, it represents, within the framework of a stadial progression that presents itself as merely incremental and progressive, an absolutely radical departure from, and incommensurability with, the limitedness that characterized the earlier stages. The final stage is not a finite stage like the others, to be succeeded and supplemented by another stage. It is truly final, because it is infinite.

And this more profound discontinuity inaugurated by the final stage also allows us to understand the logic and necessity governing the first level of discontinuity, that of stadial progression itself, which is nothing else than the logic of limits. It is precisely because the first three stages are based on means of production that are limited, that can only support a determinate level of population and material existence (although with increasing degrees of productivity), that they must eventually be overcome by another more productive stage. Indeed, it can be shown that this whole theory of stadial progression is driven by a logic of lack, what Karl Polanyi called the postulate of scarcity (Polanyi, 1968, 143), all the more pressing as one is close to an originary state of nature conceived of as inherently penurious. A conception that is not without contradictions, since it is difficult to imagine how a condition so precarious that it can scarcely support human life, such as the traditional and enduring notion of hunting and gathering, could at the same time produce the excess of population that would necessitate and enable human progression to the next, somewhat less precarious Age of Shepherds. Such, however, is the logic of limits that pervades this vision, up to and including the stage of agriculture, though this is universally recognized as the most productive of all the traditional means of subsistence, because, says

Smith, "in agriculture too nature labours along with man" (Smith, 1776, I, 363). Even so, as we have seen, agricultural production is still subject to the cycles of seasons and the limitations of natural fertility, of the finite area of cultivable land that can only sustain the growth of population up to a point – a problem that was soon to be to be posed with dramatic and disastrously misplaced urgency in the next chapter of our history of limits – and so, agriculture also must be superseded by a more productive stage. The third stage still belongs to what we might call the Age of Limits.

Only with the fourth stage, this stage which is not a stage, since it will never need to be followed by another, nor even a state, since it is defined as a self-augmenting process, will the barrier of limits be broken. The last stage is not only conceived as a terminal stage in the history of limits, but as the end of finitude itself. It is for this almost tautological reason that it must be thought as truly final. The last of all imaginable stages, because it will have no end.

From our distanced vantage point, and especially with the hindsight of the two world-changing centuries that separate us from this theoretical leap, it is tempting to be divided between several distinct, even contradictory reactions. On the one hand the rather humbling, or simply human realization that even the greatest and apparently most prescient theorists can hardly ever think beyond their own horizon. The last stage is the last simply by a common inability to imagine another stage after the present, and therefore latest stage. More exactly, the last stage is always the one that is seen to be just appearing, the stage one is "presencing" as the future.[27] No more than a projection of one's own modernity. It is hard to think of many exceptions to this, in the eighteenth century or in any other, and neither Turgot nor Smith make very good candidates. Among their near contemporaries who were serious thinkers, perhaps only Condorcet, whom we will meet in the next chapter as one of the principal foils of Malthus's first *Essay*, may have some claim to such exceptional vision, albeit not primarily as an economist.

On the other hand, it is remarkable that independently of, and well before the Hegelian account that would provide the most powerful framework for philosophies of history to come, the Four Stage Theory anticipated the idea of a universal human progression culminating in the end of history. Even if we recognize all that such a secular teleology owes to earlier theological narratives of providential history – Turgot wrote one of those discourses as a young up-and-coming Sorbonne cleric – and without laying too much stress on the resemblance between the Invisible Hand and the Cunning of Reason,[28] the last stage as it was conceived in the eighteenth century does foreshadow the theme revived in the 1980s of liberal capitalism as the End of History.[29] Although the Enlightenment economists did not yet link, as would their post-Hegelian heirs, economic liberalism with political democracy, they did prophesize the triumph of liberty in the political economy, one of the conditions for the open-endedness of the last stage.

But for the question of limits we are primarily concerned with, the most decisive step and even more fundamental condition for thinking the end of finitude is undoubtedly that the final stage, in this resolutely materialistic theory of social evolution, results in a dematerialization of the economy. This is not in the sense of the current delusion about the economy becoming virtual. The "base" for all eighteenth-century economists is still overwhelmingly agricultural, and the commerce to be promoted by the liberalization of trade is definitely that of real goods and merchandise, whether these consist in vital staples or useless luxuries. The physiocrats, with whom Turgot was first associated, even held that only the class of laborers engaged in working the land was "productive"; all others falling under the label of "sterile." While Smith thought that distinction untenable, we have seen that he too granted agricultural production a status unlike any others. But this base is that of the third stage. What characterizes the fourth stage and gives it its distinctive name is not this or any other new material mode of production in itself, but a system of exchange and investment that in principle could be applied to, and used to multiply, any productive force whatsoever. This system is not conceivable without a material basis, but it is nevertheless conceived independently of such a determinate (and therefore limited) material mode of production, and it is in this that its potential is infinite.

There are differing opinions as to whether, and how much, the early theorists of capitalism had some presentiment of the beginning of the industrial revolution.[30] Certainly both Adam Smith, as commissioner of customs in Scotland and friend to some the leading innovators and entrepreneurs of his day, and Turgot, as former *Intendant* of the Limoges region and *Contrôleur général* (Minister) of finance under Louis XVI, were in privileged positions to observe the developments under way and rapidly accelerating under their very eyes. To us, it seems obvious that they should have seen it coming. Yet we should be careful of such retrospective assurance. The manufacturing system they knew and supported was an organization of still largely artisanal production that bore little resemblance to a modern industrial plant. Its benefits were largely those of local concentration of resources and economies of scale. In giving his most famous example of the manufacture of pins at the beginning of the *Wealth of Nations*, the great principle whose advantages Adam Smith wants to extol is the division of labor, not industry, and of these advantages, the resulting improvement of machinery is only the third mentioned.[31] In fact there are only three separate passages on the benefits of machinery, and not very long ones, in the whole of his magnum opus.[32] As for the importance of energy, and especially the fossil energy that was going to be the indispensable foundation of the industrial revolution, it is hardly to be found anywhere. When he is describing the mechanical improvements that are in his time greatly cheapening and multiplying the manufacture of cloth in comparison to former centuries, a clear example of the recognition of incipient industrial transformations, Smith reminds his readers of the relatively late introduction to England of the *wind* and *water* mill (Smith, 1776, I, 263)! Of coal as the emerging source of power, not even a trace.

The greatest irony of this story – or the greatest genius of Smith and Turgot – is precisely that they dared to theorize the infinite potential of capitalism *before* the industrial and energetic revolution gave it the power to realize this potential. And in this their wager was successful beyond their wildest expectations, if only for a time. But they were right without knowing *how* they would be right. And we can now add that they were right only *because* they could not know how. For it is clear that in their universe it was unimaginable that human civilization would come to flourish on fossil energy sources on such a scale that within less than three centuries it would threaten to exhaust the entirety of the reserves accumulated through geological times, most of which were not yet discovered in their days. Even less imaginable was the prospect that the resulting explosion of human activity would one day be capable of pressing against the survivable limits of the planet itself. To eighteenth-century Europeans, apart from a few densely populated areas, most of the world they had just finally finished charting appeared almost empty, and mostly undeveloped. Its resources in land and natural productions did seem nearly infinite. And at the rate at which their exploitation could be conceived of at the time, they were indeed inexhaustible. As for those few countries whose populations could temporarily exceed local resources, we have to remember that no one in Europe, not even its most progressive thinkers and passionate critics of slavery and exploitation, had any qualms about the beneficial outlet of colonization. All the less so because the world was considered mostly empty, and the vast majority of the peoples inhabiting it were perceived to be still at the first or second stage, at most at the third. They too could begin to be brought along into the last stage, a project that could reasonably be seen as even more infinite, if at all conceivable, as we now know when it has not yet been achieved for billions of human beings, in spite of the immense economic surge of the last 200 years.

In truth we are caught in a time warp when considering the beginnings of modern capitalism. Between the theoretical models that were then developed and our own, and even more with regards to the fundamental orientation to the world that was inaugurated, we cannot help recognizing the continuity that defines our common and enduring modernity. As so many contemporary economists of various allegiances keep reminding us by retracing their lineage back to classical concepts of Smith and Turgot, it is from them and their direct followers that we have inherited not just our economics, but our most ingrained habits of thought and dispositions. But in acknowledging this foundational continuity we risk obliterating the no less profound rift in perspective between their vision and ours. For we are looking at the same prospect from opposite sides of the greatest explosion of activity in the history of humanity. Even more crucially than before industrialization, we are now discovering that they lived and thought in the era that should be called BF, before fossil fuels, whereas we are ourselves at the beginning of the third century AF, around the 200th year of the age of fossils, which is suddenly appearing as the beginning of its end, and the necessary dawn of a new era, after fossils. Of this, the founders of what is still our economic system could not possibly know or have any premonition. Their problem was not too much development and use of resources, but not enough. The irony of course is that from their model of unequal growth we have inherited the same problem as well as the opposite one. Too much and not enough at the same time, a conundrum that would have been unthinkable for those who first imagined a way out of the age of limits.

There is in *The Wealth of Nations* only one passage, or, depending on how one counts, three or even four variations on the same motif, repeated within a few pages with an almost identical core sentence, where Adam Smith does speculate on the possibility of a country which, to begin with what is apparently the most uncompromising version, "had acquired that full complement of riches which the nature of its soil and climate, and its situation with respect to other countries allowed it to acquire." Such a country, Smith's sentence concludes as a logical corollary, "could, therefore, advance no further" (ibid., 111), a statement as unequivocal as we might expect of the eventual end of growth, and surely deserving of close attention in a reflection on limits. It is true that in this first iteration the eventuality of limits to growth is expressed in the subjunctive, or conditional: a country "which had acquired" such a full complement of riches "could advance no further." And indeed, Smith seems to confirm this speculative mode by surmising a little later: "But perhaps no country has yet arrived at this degree of opulence." This hypothetical tone, however, is itself contradicted in the sentence immediately following, about China, in a formulation repeated almost exactly from the previous chapter: "China seems to have been long stationary, and had probably long ago acquired that full complement of riches which is consistent with the nature of its laws and institutions." In this case the identical core of the statement, "had acquired that full complement of riches," even modified by "probably," is no longer in the subjunctive of speculation, but in the past perfect of historical fact, just as it was in the earlier and almost identical passage: "[China] had perhaps, long before [Marco Polo's time], acquired that full complement of riches which the nature of its laws and institutions permits it to acquire" (ibid., 89). These observations about China, in conformity with the familiar and enduring view of that country as a case of stagnation or arrested progress, and in spite of the several hesitations and qualifications already noted, would seem to confirm the idea that in the course of its historical development a nation would eventually, and perhaps inevitably, reach a "full complement of riches" beyond which its wealth could no longer increase, the end of growth.

And yet with each formulation Smith also seems to offer us a kind of escape clause from the inevitability of limits. China has perhaps reached the end of growth, its greatest possible wealth in the present circumstances, but it is only the maximum "which the nature of its laws and institutions permits it to acquire," or in the second iteration "which is consistent with the nature of its laws and institutions." The ultimate limits to growth in this case would not be those of nature, or "soil and climate," but, in keeping with the philosophy of liberal economics advanced by Smith, set by the restrictive laws and institutions that prevent the optimal freedom of activity and trade on which growth depends, a suggestion clearly confirmed by the sentence immediately following: "But this complement may be much inferior, to what, with other laws and institutions, the nature of its soil, climate, and situation might admit of" (ibid., 111). And it now becomes apparent that even in the version we had found most uncompromising with natural limits, that of a country which "had acquired that full complement of riches which the nature of its soil and climate [...] allowed it to acquire," the crucial addition, and escape from inevitable finitude, is the middle clause: "and its situation with respect to other countries." For in a country's relation with other countries, its geo-political rather than its environmental condition, lies the possibility of expanding trade, and on expanded trade crucially rests the division of labor,[33] which is always for Smith the key to growth itself.

So, in the end, in spite of the traditional acknowledgment of the constraints of agriculture, and of the maximum population that a given "territory could maintain," Smith's principal concern in these developments is not with the natural limits to growth, but with the institutional and legal limitations of the capitalist system itself. Indeed, all but the earliest of the passages quoted above appear in the middle of the chapter entitled "Of the Profits of Stocks," where he imagines a time when "as great a quantity of stock would be employed in every particular branch as the nature and extent of the trade would admit." If such a point of saturation for the possibility of investment were reached, not only "the competition for employment would necessarily be so great as to reduce the wages of labour to what was barely sufficient to keep up the number of labourers" so that "that number could never be augmented," but "the ordinary rate of clear profit" (on capital) would be "as low as possible." So low, adds Smith a little later "that the usual market rate of interest which could be afforded out of it, would ... render it impossible for any but the very wealthiest people to live upon the interest of their money" (ibid., 113). As a consequence, "all people of small or middling fortunes would be obliged to superintend themselves the employment of their own stocks," and "it would be necessary that almost every man would be a man of business, or engage in some sort of trade." Here we may have the first modern adumbration of what Smith is already calling the stationary state, when the limits of economic growth would finally (inevitably?) be reached. Judging from the last sentences at least, we might be tempted to think that it would not be such a bad prospect after all. And Smith himself could be understood to lend support to this impression in the seemingly reassuring comment we have already quoted with a different emphasis: "But perhaps no country has ever yet arrived at this degree of opulence." Opulence?

We would almost certainly be wrong to interpret him so optimistically. As the earlier part of the passage makes clear, Smith, ever attentive to the effects of an inequality of conditions whose disappearance he is nevertheless unable to imagine, not even in a state of final opulence, had already warned us about the consequences of the end of growth for the majority of ordinary people:

> It deserves to be remarked, perhaps, that it is in the progressive state, while the society is advancing to the further acquisition, rather than when it has acquired its full complement of riches, that the condition of the labouring

> poor, of the great body of the people, seems to be the happiest and the most comfortable. It is hard in the stationary, and miserable in the declining state. The progressive state is in reality the cheerful and the hearty state to all the different orders of the society. The stationary is dull; the declining, melancholy.
>
> (Ibid., 99)

Thus the stage is set for a plot with which we are still inescapably familiar, in which growth is not merely desirable, but compulsory if the great majority are not to be doomed to perpetual hardship. Within such a set-up, the idea of limits will always spell misery for at least a portion of humanity. Unless we are prepared to accept such a fate, we are condemned to growth.

Of course, we are getting slightly ahead of ourselves here. If such an outcome is perfectly consistent with the logic of Adam Smith's thought, it is barely hinted at in his work in the few pages where he speculates on a distant future. For the full doomsday scenario to be spelled out, we have to turn to the next and most famous chapter in the history of limits, that of Thomas Malthus.

Notes

1 Hegel's exposition of the progressive unfolding of Spirit and Freedom through historical time is most explicit in his posthumous *Lectures on the Philosophy of History* (Hegel, 1827). Michel Foucault developed the framework of discontinuous *epistemes* and epistemological breaks that structured his earlier and best-known works in *The Archaeology of Knowledge* and *The Order of Things* (Foucault, 1976, 1973). But the notion of a fundamental break toward the end of the eighteenth century is equally prevalent in much more traditional histories of ideas such as Arthur O. Lovejoy's still wonderful *The Great Chain of Being* (Lovejoy, 1936).

2 Among the many suggestions of such an idea, certainly going back to antiquity and still present in modern philosophy, it is at least incipient, closer to the period we are concerned with, in René Descartes' *Meditations on First Philosophy* (Descartes, 1641) and Hegel's *Lectures on the Philosophy of Religion* (Hegel, 1827), but these are only possible openings for a large investigation.

3 For serious studies on the subsistence of hunting and gathering societies, see for example *Man the Hunter* (Lee and De Vore, 1968), particularly Richard B. Lee's "The Hunters: Scarce resources in the Kalahari," as well as Allen Johnson's "In Search of the Affluent Society" (Johnson, 1978) and Marshall Sahlins' *Stone Age Economics* (Sahlins, 1972). For an older and perhaps more guarded view, Melville J. Herskovits' *Economic Anthropology*, Chapter 5 (Herskovits, 1960).

4 According to Rousseau's conjectural and highly idiosyncratic stage theory, the "first revolution," which corresponds to the first socialization of originally dispersed humans, is brought about by the simultaneous invention of agriculture and metallurgy. *Discourse on the Origins and Foundations of Inequality among Men*, Part Two (Rousseau, 1754).

5 For a thought-provoking account of this process, see Richard Manning: *Against The Grain: How Agriculture Hijacked Civilization* (Manning, 2004).

6 There are many translations, or renderings, necessarily interpretive, from the fragmentary remains of the cuneiform tablets on which was preserved what we call *The Epic of Gilgamesh*. My favorite is by N. K. Sandars, with a substantial "Introduction" by the translator (Sandars, 1960).

7 See the wonderfully readable *Collapse*, by Jared Diamond (Diamond, 2011), and the earlier and more demanding but equally rewarding *The Collapse of Complex Societies* by Joseph Tainter (Tainter, 1988).
8 Edward Gibbon: *The History of the Decline and Fall of the Roman Empire*, published between 1776 and 1789 (Gibbon, 1776–1789).
9 For conceptions of time in Antiquity, see Arthur O. Lovejoy and George Boas: *Primitivism and Related Ideas in Antiquity* (Lovejoy and Boas, 1935), and for their ongoing influence through the eighteenth century, the still untranslated *Anthropologie et histoire au siècle des Lumières*, by Michèle Duchet (Duchet, 1977).
10 Montesquieu's magnum opus, *The Spirit of the Laws*, was first published in French in 1748, and translated into English in 1750 (Montesquieu, 1750). It was arguably one of the most influential books in the eighteenth century, certainly one of the most widely studied across the European world.
11 At mid-century, the lowest percentage of the population engaged in agriculture was in England, at 65 percent. Sweden was at 75 percent and France at 76 percent, which probably represents a low European average. From Carlo Cipolla: *Before the Industrial Revolution* (Cipolla, 1980, Chapter 2).
12 The most important member and founding father of the group of eighteenth-century economists who called themselves the Physiocrats was François Quesnay, author of *Physiocracy*, and best known as the inventor of the *Tableau Economique* (Economical table) (Vaggi, 1987). Turgot, whom we will soon acknowledge for his early formulation of the Four Stage Theory, was also associated with the Physiocratic movement and agreed with some of its tenets, notably its advocacy of free trade in grains.
13 For technological progress and pre-industrial organizations in the late Middle-Ages, see Jean Gimpel: *Medieval Machine* (Gimpel, 1977) and Lynn White, Jr.: *Medieval Technology and Social Change* (White, 1962).
14 This is also not quite true, since coal, called "pit coal" or "sea coal" as opposed to charcoal began to be used in England and France as early as the thirteenth century, already in the lime and iron industries, but also for heat, in such quantities and poor quality as to create significant pollution in cities. See Jean Gimpel, op. cit. (Gimpel, 1977, 81–84).
15 See in particular Carol Blum: *Strength in Numbers*, Chapter 2 "Montesquieu and the 'Depopulation Letters'" (Blum, 2002); Robert J. Mayhew: *Malthus* (Mayhew, 2014, 20–24), and more generally Charles Emil Stangeland: *Pre-Malthusian Doctrines of Population* (Stangeland, 1966).
16 In part because he was so active as an administrator and later minister under Louis XVI, Turgot (Anne Robert Jacques) did not leave a body of work as complete and ultimately influential as Adam Smith, and he is today remembered mostly in France as a historical figure, and by students of economic thought. But among the latter he is a well-respected exponent of early economic liberalism, and one of the first to anticipate the law of diminishing returns in agriculture. Joseph Schumpeter devotes no less than seven rather laudatory pages to him (Schumpeter, 1954, 243–249).
17 The expression appears in Michel Foucault's essay "What is an Author?" (Foucault, 1984).
18 In English, as it were, commonly used in the imperative mode of *laissez-faire*. Although there is some doubt as to who used it first in this form, the maxim was widely adopted by the aforementioned Physiocrats, and from them by the young Turgot. The philosophy of economic liberalism, however, was by no means exclusively theirs, nor uniquely French, being espoused by Adam Smith and David Hume before him.
19 The classic exposition of this major shift in the moral status of self-interest is Albert O. Hirschman's *The Passions and the Interests* (Hirschman, 1977).
20 The deceitfulness of *amour propre*, or self-love, and the sway of vanity over human passions, are the dominant themes of La Rochefoucauld's *Maxims* (1665), a

collection of clever and elegant aphorisms. Maxim number 2 provides perhaps the most concise summary: "Self-love is the greatest flatterer of all" (La Rochefoucauld, 1665, 37).

21 Jansenism, named after the Dutch theologian Cornelius Jansen, was a movement of austere piety and morality within the Catholic church, espousing a deeply pessimistic view of human nature and a conception of grace (salvation) close to that of Calvinism in the Protestant tradition. The philosopher Pascal, evoked in our first chapter, was close to the Jansenists and a sometime resident at their spiritual retreat in the Abbey of Port Royal near Paris.

22 Turgot's terms for the first three stages are: first "*Chasseurs*," second "*Pasteurs*" and third "*Laboureurs*" (husbandmen, literally ploughmen). *Plan d'un ouvrage sur la géographie politique*, and *Plan de deux discours sur l'histoire universelle* (Turgot, 2018, I, 259–253, 277–282).

23 A law Turgot himself had anticipated with regards to agriculture. As we shall see below, even without such a concept, Adam Smith has his own reservations about the extent to which the productivity of the soil could be increased. The factor called "land," when it is not ignored, is always the most subject to limits in a long-term theory of growth.

24 Let us remember that *Natura non facit saltum*. In most traditional conceptions of the world, such as the "Great chain of beings," nature does not make a jump – with the exception perhaps of some theories of evolution. A stadial theory of historical (human) progress, on the contrary, "jumps" from one stage to the next.

25 The "stages" in Giambattista Vico's cyclical vision of history, exposed in *Scienza Nuova* of 1725, the divine, the heroic, and the human, are evidently not materially defined (Vico, 2000). As for Bossuet's *Discourse on Universal History* (1681), which must have been a cornerstone of Turgot's education and a model for his own early sketches of universal stages of human progress, it is explicitly theological (Bossuet, 1681).

26 This point was raised quite explicitly, if not in exactly the same terms, by Istvan Hont in "The Language of Sociability and Commerce: Samuel Pufendorf and the Theoretical Foundations of the 'Four-Stage Theory'" (Pagden, 1987, 254).

27 Michel Foucault, a later day stage theorist who was not immune to the temptation to prophesize, nevertheless had these wise words to say about the illusion:

> I think we should have the modesty to say to ourselves that [...] the time we live in is not *the* unique or fundamental or irruptive point in history when everything is completed and begun again [...] It is a time like any other, or rather, a time that is never quite like any other.
>
> (Foucault, 1998, 449)

28 In Hegel's Philosophy of History, the cunning of reason designates the process whereby Reason accomplishes itself in history through the agency of great men unconsciously embodying successive national spirits, each progressively more perfect than the previous one.

29 Notably in Francis Fukuyama's *The End of History and The Last Man* (Fukuyama, 1992).

30 With regards to Adam Smith, the prevailing view among scholars is that, although he was a contemporary and in some cases even a neighbor of inventors and early industrialists such as Richard Arkwright, Matthew Boulton and James Watt, his observations did not extend much further than small-scale manufacturing. See Hiram Caton: "The Preindustrial Economy of Adam Smith" (Caton, 1985); R. Koebner: "Adam Smith and the Industrial Revolution" (Koebner, 1959) and C. P. Kindleberger: "The Historical Background: Adam Smith and the Industrial Revolution" (Wilson and Skinner, 1976, Chapter 1).

31 See Chapter I: "Of the Division of Labour" (Smith, 1776, 14–20).

32 Without disagreeing about the relatively low level of prescience demonstrated by Smith about the coming industrial revolution, W. W. Rostow is more generous about his awareness of the longer-trend effects of technological innovation on economic growth:

> [Smith] did recognize, however, that over long periods of time one could identify a few major technological innovations whose high productivity constituted an identifiable discontinuity helping to explain a long period decline in price [for example the spinning wheel and the fulling mill in the woolen industry].
>
> (Rostow, 1990, 41)

33 See Chapter III: "That the Division of Labour is limited by the Extent of the Market" (Smith, 1776, 31–36).

References

Blum, Carol (2002) *Strength in Numbers: Population, Reproduction, and Power in Eighteenth-Century France*. Baltimore and London: The Johns Hopkins University Press.

Bossuet, Jacques Bénigne (1681) *A Universal History from the Beginning of the World to the Empire of Charlemagne*. Andesite Press (this edition 2017).

Caton, Hiram (1985) "The Preindustrial Economy of Adam Smith," *Journal of Economic History*, vol. 45, no. 4, 833–853.

Cipolla, Carlo (1980) *Before the Revolution: European Society and Economy 1000–1700*. New York: W. W. Norton.

Descartes, René (1641) *Meditations on First Philosophy*. Indianapolis and Cambridge: Hackett Publishing Company (this edition 1993).

Diamond, Jared (2011) *Collapse: How Societies Choose to Fail or Succeed*. New York: Penguin.

Duchet, Michèle (1977) *Anthropologie et histoire au siècle des Lumières*. Paris: Flammarion.

Foucault, Michel (1973) *The Order of Things: An Archaeology of the Human Sciences*. New York: Vintage Books.

Foucault, Michel (1976) *The Archaeology of Knowledge*. New York: Harper and Row.

Foucault, Michel (1984) "What is an Author?" in *The Foucault Reader*, Paul Rabinow (ed.). New York: Pantheon Books.

Foucault, Michel (1998) "Structuralism and Post-Structuralism," in *Aesthetic, Method, and Epistemology*, James D. Faubion (ed.). New York: The New Press.

Fukuyama, Francis (1992) *The End of History and The Last Man*. New York: Avon Books.

Gibbon, Edward (1776–1789) *The History of the Decline and Fall of the Roman Empire*. London: Penguin Books.

Gimpel, Jean (1977) *Medieval Machine: The Industrial Revolution of the Middle Ages*. London: Penguin Books.

Hegel, G. W. F. (1827) *Lectures on the Philosophy of Religion*. Berkeley, CA, Los Angeles and London: University of California Press (this edition 1988).

Hegel, G. W. F. (1837) *Lectures on the Philosophy of History*. Aalten, The Netherlands: Wordbridge Publishing (this edition 2011).

Herskovits, Melville J. (1960) *Economic Anthropology: A Study in Comparative Economics*. New York: Alfred A. Knopf.

Hirschman, Albert O. (1977) *The Passions and the Interests: Political Arguments for Capitalism before its Triumph*. Princeton, NJ: Princeton University Press.

Johnson, Allen (1978) "In Search of the Affluent Society," *Human Nature*, vol. 9, no. 1, 50–59.

Koebner, R. (1959) "Adam Smith and the Industrial Revolution," *Economic History Review*, vol. 11, no. 3, 381–391.
La Rochefoucauld, F. (1665) *Maxims*. London: Penguin Books (this edition 1959).
Lee, Richard B. and De Vore, Irven (eds.) (1968) *Man the Hunter*. New York: Aldine Publishing Company.
Lovejoy, Arthur O. (1936) *The Great Chain of Being: A Study of the History of an Idea*. Cambridge, MA: Harvard University Press.
Lovejoy, Arthur O. and George Boas (1935) *Primitivism and Related Ideas in Antiquity*. Baltimore: Johns Hopkins University Press.
Manning, Richard (2004) *Against the Grain: How Agriculture Hijacked Civilization*. New York: North Point Press.
Mayhew, Robert J. (2014) *Malthus: The Life and Legacies of an Unlikely Prophet*. Cambridge and London: Cambridge University Press.
Meek, Ronald L. (1976) *Social Science and the Ignoble Savage*. Cambridge, UK: Cambridge University Press.
Montesquieu (1750) *The Spirit of The Laws*. Cambridge, UK: Cambridge University Press (this edition 1989).
Pagden, Anthony (ed.) (1987) *The Languages of Political Theory in Early Modern Europe*. Cambridge, UK: Cambridge University Press.
Polanyi, Karl (1968) *Primitive, Archaic and Modern Economies: Essays of Karl Polanyi*. Garden City: Anchor.
Rostow, W. W. (1990) *Theorists of Economic Growth from David Hume to the Present: With a Perspective on the Next Century*. New York and Oxford: Oxford University Press.
Rousseau, Jean-Jacques (1754) *A Discourse on Inequality*. Harmondsworth, Middlesex: Penguin Books.
Sahlins, Marshall (1972) *Stone Age Economics*. New York: Aldine de Gruyters.
Sandars, N. K. (1960) *The Epic of Gilgamesh*. London: Penguin Books.
Schumpeter, Joseph A. (1954) *History of Economic Analysis*. New York: Oxford University Press.
Smith, Adam (1759) *The Theory of Moral Sentiments*. Oxford: Oxford University Press (this edition 1976).
Smith, Adam (1776) *An Inquiry into the Nature and Causes of the Wealth of Nations*. Oxford: Oxford University Press (this edition 1976).
Stangeland, Charles Emil (1966) *Pre-Malthusian Doctrines of Population: A Study in the History of Economic Theory*. New York: Augustus M. Kelley.
Tainter, Joseph (1988) *The Collapse of Complex Societies*. Cambridge, UK: Cambridge University Press.
Turgot, Anne Robert Jacques (2018) *Oeuvres de Turgot, et documents le concernant*, Gustave Schelle (éd.). Paris: Institut Coppet.
Vaggi, Gianni (1987) *The Economics of François Quesnay*. Durham: Duke University Press.
Vico, Giambattista (2000) *New Science*. London: Penguin Classics.
White, Lynn, Jr. (1962) *Medieval Technology and Social Change*. Oxford: Oxford University Press.
Wilson, Thomas and Andrew S. Skinner (eds.) (1976) *The Market and The State: Essays in Honour of Adam Smith*. Oxford: Oxford University Press.

3 The Malthusian moment

If one were to devise a ratio contrasting an author's fame, particularly measured by name-recognition, with how well his or her work is really known, or how little it is actually read, Malthus, whose reputation famously rests on the promulgation of a provocative ratio, may well rank at the top of such a differential scale. Among the truly great contenders, perhaps only Darwin comes close, although his ideas are better known and almost universally recognized. Marx, whose influence and recognition are of course much greater, was also read and proselytized to an incomparably wider extent, even if it is less so nowadays. At any rate very few thinkers of any comparative caliber have been honored by posterity with both a common adjective coined after their name – even if in Malthus' case the adjective is most often used as an epithet – and a substantive uniquely denoting the concept he promoted, that of inescapable limits to population, and even with a secondary current of thought known as neo-Malthusianism, which stands for the advocacy of birth control, much as Malthus himself disapproved of the idea.[1] For this reason alone, Malthusianism, if not its originator, would have to figure prominently in any reflection on limits, as no proper name is more closely associated with the question we are pursuing.

A second reason, or perhaps another aspect of the same contrast, is that in spite of his largely negative reputation, Malthus has had over time, among those who did read him, many professed admirers and defenders. And not least among those who were certainly qualified to understand him, some of the greatest economists the world has ever known: John Stuart Mill, Alfred Marshall, and John Maynard Keynes. And to touch already on one of the dominant themes of this chapter, none of these celebrated figures belongs in the conservative moral and political camp with which we usually, rightly or wrongly, associate Malthus. They are on the contrary widely considered to be socially progressive and imminently humane representatives of the science said to be dismal. But then, even on the very level of personality from which Malthus' reputation has also suffered, those who have studied him most closely frequently vouch that he was privately a man of kind disposition, and, true to his Christian calling, sincerely concerned with the welfare of humanity. And yet – this is the final twist in this series of contrasts – even the most sympathetic among his scholars[2] must recognize that Malthus, and the various labels associated with his name, cast in the history of human progress an essentially negative, even ominous spell.

For beyond name-calling, and occasionally because of it, Malthus in fact did much to provoke and influence the question of limits, both deliberately in the way he posed it, and indirectly through the many reactions, often passionate and as often hostile, that his position has elicited from the publication of his first *Essay* almost to our own time. Thus Malthus' impact is not only important historically; we will see that it also contributed more than any other to framing the subsequent debate on the necessity or possibility of limits, and set patterns of reaction that are still reverberating, most forcefully in opposition to his own pronouncements.

But first an obvious objection to the line of thought we have been following up to this point will surely come to mind. If the most resounding proclamation of the inexorability of limits occurred so soon after the triumphant late-Enlightenment declarations of economic independence we have just left – the first version of Malthus' *Essay on the Principle of Population* is dated 1798, not even yet over the brink of the next century – doesn't this call into question the idea of a fundamental break in the history of limits, of a turning point initiated by the great theorizations of the fourth stage of economic progress. After all, both Malthus, in his own right a major classical economist, and the other even more important "pessimist" of classical economy, David Ricardo, claimed for the most part to be disciples of Adam Smith, who as we last saw was not by temperament of a particularly optimistic bent. What appeared to us as a major discontinuity may be more accurately understood as a shorter fluctuation within ongoing trends, or a pattern of generational reactions between conflicting currents of thoughts. This objection must be taken seriously. Aside from the caution that should attend any attempt to describe large patterns in the history of thought, the perception of great discontinuities can only be achieved at the expense of fully recognizing the complexity, not to say messiness, of lasting influences, of predecessors and holdovers that are no less true. And the impact of individual thinkers, of which Malthus is a singular example. His notoriety could signal a return, or even a persistence of the idea of limits.

It is mostly just the opposite. The scandal and hostility provoked by the *Essay*, in spite of the fact that in this first iteration at least, it arguably contained no major idea or evidence that had not been in some form advanced before,[3] except perhaps for the demonstrative use of ratios, reveals that by the time of its publication a page had definitely been turned on the inevitability of limits. What had only a half-century before been generally accepted as common sense – that we live in a limited world – had by Malthus' time become an unacceptable proposition, a provocation against the new horizon of progressive expectations. It is true that Malthus did much to fuel this sense of provocation by deliberately taking on as his target two well-known progressive thinkers of his days, as well as by his rhetoric. But most of all, in the history of limits, his timing was wrong. Like a Don Quixote of limits, he had stumbled into the new era of progress as the last somber knight of finitude,[4] steadfastly upholding the older certitude of doom for the laboring multitude, as well as the social order that sustained it, just when the possibilities for an unprecedented improvement of the living conditions of the great majority of mankind were being opened, and beginning to be felt.

And this allows us to understand the two main reasons why, in the history we are pursuing, the legacy of the most notorious champion of limits is so overwhelmingly negative: he turned out to be fatally wrong in his mid to long-term predictions, even though his fundamental argument was in some basic sense right, and he managed to alienate, for a very long time to come and in many ways up until our own day, most of the progressive currents of thought in what we have come to know as the social and political Left. Such a historical verdict undoubtedly fails to do justice to the complexity of Malthus' actual work, and certainly to his intentions, since there is no reason to question his professions of benevolence,[5] but for the modern destiny of the notion of limits, these are the two aspects of his legacy that have proved truly momentous.

The famous ratios are nothing but a pedagogical device, meant to impress to be sure, but in the end, rather harmless. Malthus begins with the observation that in the unusually favorable conditions of colonial expansion in North America, settler populations had been known to double in 25 years, or one generation. Hence his choice for the exponential rate of natural population increase, absent limitations. It is admittedly arbitrary. In a later chapter on Godwin, Malthus mentions, "on the authority of a pamphlet published by a Dr. Styles" referred to by Richard Price, "that the inhabitants of the back woods of America doubled their numbers in fifteen years" (Malthus, 1798, 198), but just to be sure, he remains with the more conservative estimate, no doubt for its ease of computation. His choice for the possible rate of increase in food production, a doubling during the same first 25 years, but then increasing only by the same absolute amount every subsequent 25-year period, an "arithmetical ratio," is even more arbitrary, although by Malthus' own logic it must have been initially achieved by his American settlers. Based on the knowledge and practices of his own time, imagining such a sustained rate of improvement in agricultural production was indeed optimistic, and so seems a generous choice on his part. We now know that comparable increases were in fact achieved and even temporarily exceeded in the twentieth century by using all the scientific advances and massive mineral boost (and public subsidies) of the "green revolution," albeit at a price we are only now beginning to recognize, but Malthus certainly cannot be faulted for thinking that he was granting his critics every conceivable margin of doubt. Even so, it is easy to see how by extrapolating over several generations such differential ratios of population and subsistence growth, one quickly arrives at an apocalyptic scenario. "In two centuries and a quarter," Malthus projects (coincidentally coming fairly close to our own present time), "the population would be to the means of subsistence as 512 to 10: in three centuries as 4096 to 13; and in two thousand years the difference would be almost incalculable, though the produce in that time would have increased to an immense extent" (ibid., 25–26).

It would be. But of course it won't, because according to Malthus' own sound principle, population cannot increase faster than the means of subsistence necessary to sustain it, so that the threatened exponential growth of disparity between the two would end promptly after the first 25-year period.[6] The argument of the ratios is nothing more than a theoretical scare tactic, and even if it did seem frightening enough to have some lasting resonance, it was soon seen (even by Malthus himself, although he never disowned it) to be rather meaningless.

On the empirical side, the evolution of the argument is noticeably different, but in the end perhaps equally tautological. In the first version of the *Essay*, Malthus had devoted two early chapters to a quick historical sketch of the operation of constraints on population – what he calls "checks" – in various types of societies, following a compressed version of the by now familiar framework of the Four Stage theory, from the "savage or hunter state" and the "shepherd state" in Chapter 3, to the "state of civilized nations" in Chapter 4. After the publication of this first edition, rather hurried by his own admission, Malthus evidently felt that he needed to provide corroborating evidence for his theory, and engaged in an extensive study of available ethnographic and historical documents, ranging over a wide span of time and geographical array of the known world, supplemented by his own fact-finding travels in Scandinavia, Russia, France and Switzerland.[7] The result of this impressive effort is the second edition of the *Essay* in 1803, so enormously enlarged from the first as to be barely recognizable. The original sections have been mostly retained, but in an unedited version they are now nearly overwhelmed by long and numerous chapters of supporting documentation. Only to what effect? Predictably, Malthus finds everywhere and at all times confirmation that the numbers of human populations have been constrained by the limits of food production, and if not checked in their birthrates, then kept down by famine, epidemics, wars, or forced emigration. Given his avowed intentions in undertaking this research and the premise he was setting out to verify, it is difficult to imagine how it could have been otherwise. Indeed, in the Preface to the 1803 edition, Malthus candidly recognizes: "In drawing similar inferences from the state of society in a number of different countries, I found it very difficult to avoid some repetitions" (Malthus, 1803, 8). There are moments, particularly when attempting to account for ethnographic observations, when he does admit to not fully understanding why some populations had not increased as fast as should have been expected, but this did not lead him to revise any of his conclusions, quite literally forgone in this case, as the *in extenso* reproduction of the 1798 opening chapters confirms. In the end, this whole labor of *post-theoriam* verification, extensive as it is, feels increasingly redundant, because it is. Population cannot grow faster than the means of subsistence. If that were all that Malthus had wanted to prove, as he seems to do in these additional chapters, he would probably not have been any more noticed, nor become more notorious, than his obscure predecessors.

But the real issue with limits to subsistence is not to recognize their existence, although that is a necessary first step, but to know what can be done about them, either by extending those limits, or by changing human behavior or social organization so as to learn to live within them with the least amount of suffering. As is well known, it is in his response to these questions that Malthus was most profoundly negative, and at least in the first editions of the *Essay*, flaunting a degree of pessimism about the human condition that approaches callousness.[8]

That he was not more prescient about the prospect of increasing food production is in itself unexceptional. As already noted, his allowing for an arithmetic increase of the means of subsistence, whether he really believed it possible or not, was actually optimistic for the time. And it is worth stressing that Malthus's exclusive concern in the *Essay* is with the limits of food. He did not envisage such a limitation to industrial production, insisting on the contrary that the latter was infinitely expandable. "It should be remembered always," he writes in the chapter on positive checks to population, "that there is an essential difference between food, and those wrought commodities, the raw materials of which are in great plenty. A demand for these last will not fail to create them in as great a quantity as they are wanted" (Malthus, 1798, 90). This distinction is of great importance for the way the question of population is posed, both in Malthus' time and in our own. Now that we have learned that there are also limits to the production of "wrought commodities," we realize that the footprint of the relatively few rich inhabitants of the planet may be as destructive as the survival needs of the many more poor ones, a consciousness that has recently pitted the sparsely peopled but high-consuming North against the overpopulated and still hungry South, but might still lead to the realization that both threats must be jointly addressed to achieve an equitable sharing of the earth. For Malthus, the population problem is strictly a matter of providing a minimum of living sustenance for the ever-pressing masses of the laboring poor. He is aware that this bare minimum is subject to some variation "arising from the different habits of living which prevail in each state," offering for example the faintly disapproving illustration that: "The labourers of the south of England are so accustomed to eat fine wheaten bread, that they will suffer themselves to be half-starved before they will submit to live like the Scotch peasants" (Malthus, 1803, 41). But there is no doubt that for him, the pressure of population on the standard of living is ever downward, and from this pressure, no significant relief is to be looked for in the prospect of industrial progress, infinite as that may be.[9]

For the most crucial consequence of the "essential difference" is a failure to recognize that the "wrought commodities" and even the "raw materials of which are in great plenty" could be used to increase the production of food itself. Malthus sensibly insists that no amount of wealth created by the manufacturing system will sustain more laborers unless it can be converted into a greater agricultural product (Malthus, 1798, 306), but he evidently cannot foresee how industrial processes could be applied to that effect, either through mechanization, or the manufacture of non-animal-based fertilizers. This is made quite clear by another precious illustration, when Malthus is trying to imagine how the total produce of England could possibly be doubled in 25 years (his stated optimum ratio): "The only chance of success would be the ploughing up all the grazing countries, and putting an end almost entirely to the use of animal food." But then he is compelled to infer: "Yet a part of this scheme might defeat itself. The soil of England will not produce much without dressing; and cattle seem to be necessary to make that species of manure, which best suits the land" (ibid., 187). The reasoning is sound, and on technical grounds alone, Malthus cannot be faulted for simply reflecting the known practices of his own time. But in retrospect, we can see that his is a spectacular case of failure of imagination, if not of refusal to admit the possibility of as yet undiscovered applications to the production of food. And in this he is not merely reflecting the conceptual possibilities of the period, but appears to be on the low side of its horizon of expectations. By contrast, and in anticipation of the debate we will soon follow in some detail, here is Condorcet – to be sure, in a mode as visionary as Malthus' was pessimistic – looking forward, a few years *before* the Essay was written, to a future time when agriculture would benefit from all the advances of scientific and technological progress:

> A smaller and smaller area of ground will then be able to produce an amount of foodstuff of greater utility or higher value; more extensive enjoyment will be obtained with less consumption [of resources]; the same product of human industry will require a lesser destruction of raw materials, or will remain useful for a longer period of time. We will be able to choose, for each soil, those productions that fulfill more numerous needs, [or] between productions that can satisfy needs of the same kind, those that satisfy a greater number, while requiring less real consumption. Thus, without any sacrifice, the means of conservation, of economy in consumption, will follow the progress of the art of reproducing diverse materials, of preparing them, and of using them to manufacture different products.[10]
>
> (Condorcet, 1795, 280–281)

The passage is all the more extraordinary as the older Condorcet is not only anticipating industrial developments that would have been difficult to predict in his (and Malthus') time, but even, beyond this era of progress in mass production and consumption, the need to preserve resources and use them more efficiently, evoking such strikingly modern concepts as durability and conservation, if not recycling. It would almost seem unfair to compare Malthus' prognostic with such an exceptional case of clairvoyance, except for the fact that this is one of the two texts that the prophet of doom specifically singles out in order to discredit them as idealistic fantasy.

Whether Condorcet can be said to have been vindicated by history, or just pithily inspired in a moment of enthusiastic vision, it is certain that Malthus was not. And in the business of forecasting, there is nothing worse for the fate of the prophecy than for the prophet to have been proved wrong. This is the first lesson of the Malthusian moment for the history of limits. It almost does not matter that Malthus had understandable reasons for underestimating the potential of the industrial revolution – in the same way that we saw his immediate forebears largely unconscious of its imminent coming – and that he was in fact pointing to a very real long-term problem. For the next 200 years after he made it, surely a time-span sufficient to verify a historical hypothesis, his prediction visibly failed to be fulfilled. It is true that millions of people throughout what was once known as the Third World continued to suffer from endemic malnutrition and periodic famine, as they still do to this day for a range of reasons, but in the industrializing countries Malthus was clearly focusing on, the opposite happened. Not only did food keep pace with increasing population, but the amounts and richness of nutrients available per capita increased enormously, so much so that affluent populations are now encountering problems associated with overfeeding. It is also true, as has already been signaled a number of times, that this was achieved by methods and use of non-renewable resources that are proving unsustainable and hugely destructive as well as humanly exploitative, so that the problem of feeding the planet's soon to be eight billion inhabitants is by no means resolved. Nevertheless, in the period during which Malthus' dire prediction has so far been assessed, he has justifiably appeared to be widely off the mark in warning of impending disaster.

Thus the first and best-known meaning of Malthusianism. To be Malthusian is to cry wolf about limits, generally because one fails to estimate the human capacity to postpone those limits through ingenuity and unforeseen scientific discoveries. It has had a long and very damaging legacy for the recognition of the importance of limits, obviously working as a counter-effect to the best intentions of those arguing most sincerely and passionately for that recognition. Much closer to us, we can cite as famous examples the untimely warnings (though not untimely forever) of Paul Ehrlich and of the so-called Club of Rome *Limits to Growth* (Ehrlich, 1968; Meadows *et al.*, 1972). For the many more who are trying to engage in a thoughtful but no less urgent reflection on the significance of limits, it is a lesson in the need to exercise great caution about making predictions, and for being especially attentive to the arguments of those who differ from us, lest we repeat the same mistakes. For fear, at least, that we be called Malthusian.

The second meaning, perhaps less clearly recognized though no less momentous, is more political. Political in the sense that Malthusianism, understood as the idea that there are natural limits to human expansion and betterment that will prove intractable to social progress, is strongly associated, or at least was so for a very long time, with the right, or conservative side of the political spectrum. This is of course an oversimplification, and to do full justice to the many questions it raises is beyond our reach, but we must at least first recognize that such association of an idea with one or the other side of the unstable Right/Left topology inherited from the history of Western politics is itself historical, and liable to change. Some notions or ideals that originated on one side have been known to migrate to the other, or to become neutral in a once polarized field. Nationalism, for example, now mostly owned by the political right (in the West, though not yet everywhere), was originally born in the revolutionary movements we normally associate with the Left. To situate free market capitalism in such a polarity is by no means as easy as it now appears. Liberalism, as the word indicates, originated in a wider movement of liberation from the social constraints of the *Ancien Régime*; in the eyes of the Oeconomists, those of the old system of jurands and corporations and the arcane rules that restricted the production and trade of goods, and therefore the wealth of nations for the benefit of all. The forefathers of capitalism, such as the Physiocrats, were in this sense progressive, even if they remained hesitant to embrace the ideals of the democratic left. Turgot, their erstwhile disciple and Condorcet's master, was not far from being considered radical in his reformist ambitions, minister to the king though he was for a short period. Even the more sober and temperamentally moderate Adam Smith was by no means viewed as a conservative. Only with the "pessimist" economists of the early nineteenth century, such as Ricardo and Malthus himself, do we begin to see a strong association of liberal economics with the political right. But we know very well that not every economist who has since then embraced the benefits of market capitalism has such leanings. If one were really honest about it (though this is something left-wing politicians still find it difficult to admit, especially in Europe and Latin America), one would recognize that some moderate version of capitalism has now become politically neutral. Ironically, only hard-core economic "liberalism" is now militantly "conservative."

The location of the environmental movement – the other large conceptual field with which our question intersects – in relation to this same political spectrum is perhaps even more complex. Although contemporary Green activists and political parties are decidedly on the Left, the thinkers and writers who first called attention to the value of preserving natural environments, and are therefore the historical founders of the conservationist movement (conservatives in another derivation of the term) hailed from both sides of the Left–Right divide, or would be difficult to place. In the United States it should be remembered that the ambitious programs of wilderness set-asides that created the National Parks, nothing less than visionary in retrospect, were initiated under Republican administrations, a fact that the undisguised hostility to environmentalism of more recent leaders of the same party might make us forget. Ecologists are not naturally leftists, though many would be tempted to think that they are, and conversely, environmentalists, especially if they are from the West, are often looked upon with suspicion by left-leaning politicians from the South. If they forefront the threat of human overpopulation to wildlife survival, they are likely to be branded as privileged elitists, if not much worse. All these examples are not meant to exhaust the question, only to suggest that the political location of a preoccupation with limits was not from the start a foregone conclusion, and that it has again been very much in flux in more recent times.

This makes Malthus' historical role in anchoring this location on the Right even more visible and pivotal, with far-reaching repercussions whose effects are not yet exhausted.[11] And the attribution of his insistence on the crushing weight of limits to a conservative position is not a misunderstanding on the part of his coeval or later commentators.[12] For Malthus himself, it is explicitly grounded in an anthropology, which is itself founded on a theological interpretation exposed in the last two chapters of the *Essay*. According to this view, the mass of mankind is so naturally indolent and slothful that it would have remained in its original brutish condition if not spurred by the pressure of survival needs. The natural imbalance between our reproductive capacity and our ability to procure resources is thus the way Providence rouses us into action:

> if we return to the principle of population, and consider man as he really is, inert, sluggish, and averse from labour, unless compelled by necessity, [...] we may pronounce, with certainty, that the world would not have been peopled, but for the superiority of the power of population to the means of subsistence.
>
> (Malthus, 1798, 363–364)

Furthermore, without enduring hardship as the condition of the ever-breeding multitudes, humanity would fall back to its primal torpor. The misery of the greater numbers is not just an evil impossible to eradicate, but also a necessary moral spring to save us from even worse misery. "Necessity has been with great truth called the mother of invention" (ibid., 358). Looked upon from a higher prospect, overpopulation is a beneficial curse.

Malthus is of course well aware that not all human beings are subject to the same iron law, since quite clearly the middle and upper classes of society manage to rise above it through restrictions to marriage and "restraint." But on the one hand he does not seem to believe the great majority of mankind to be capable of such self-control, and on the other their social betters do not need the same basic incitement of survival pressure to be induced into the pursuit of civilizing activity. In their case the pressure to keep up with one's social rank and station performs the same function. And this social order is in turn naturalized by an organic analogy: "The middle regions of society seem to be best suited to intellectual improvement; but it is contrary to the analogy of all nature, to expect that the whole of society can be a middle region" (ibid., 367). Just as

> in the oak, the roots and branches could not be diminished very greatly, without weakening the vigorous circulation of the sap in the stem; so in society, the extreme parts [of riches and poverty] could not be diminished beyond a certain degree, without lessening that animated exertion throughout the middle parts.

For "if no man could hope to rise, or fear to fall, in society; if industry did not bring with it its reward, and idleness its punishment, the middle parts would not certainly be what they are" (ibid., 369). Thus Malthus manages to justify both the misery of the laboring masses and the permanence of social inequality through the principle of population. Limits at once naturalize the necessity of a hierarchy of wealth among men, and the inevitability of suffering for those at the bottom.[13]

That this is not the only possible way to think about limits is one of the points the present reflection would most dearly wish to make, but it is also an argument that Malthus was already well aware of in his own post-revolutionary times.[14] In order for his position to hold, he must disprove the hopes that humanity can change, either in terms of moral and material progress, or of its unequal social arrangements. It is therefore no wonder that the *Essay*'s most polemical chapters are devoted to a refutation of two works that earnestly argue for the possibility of human perfectibility and equality, William Godwin's *Enquiry Concerning Political Justice, and its Influence on Morals and Happiness* (1793), and Condorcet's *Esquisse d'un tableau historique des progrès de l'esprit humain* (*Outlines of an historical View of the Progress of the Human Mind*), posthumously published in 1795, but written around the same time as Godwin's essay.

Of the two, the former, which according to Malthus provided the original impetus for writing his *Essay*, is by far the easier target. Godwin is unabashedly writing in a utopian mode, and he is the woollier thinker. His views on the human predicament and the way to resolve it are also precisely the opposite of Malthus' naturalism, "attributing almost all the vices and miseries that are seen in civil society to human institutions" (Malthus, 1798, 176). And so, when Godwin dreams of social arrangements that would guarantee equal plenty and perfect liberty for all by abolishing private property and the "monopoly" of marriage, Malthus has a field day dashing any hope that such an ideal community could survive the test of reality, much as Mandeville had done in *The Fable of the Bees*. Indeed, the combination of freedom from hardship provided by the collective sharing of resources and total liberty in the "commerce of the sexes" would not fail to produce, in his perspective, the perfect recipe for disaster. "I cannot conceive," he logically concludes, "a form of society so favourable upon the whole to population" (Malthus, 1798, 184). Instead of Godwin's hope that in his selfless world "myriads of centuries of still increasing population may pass away, and the earth be still found sufficient for the subsistence of its inhabitants" (ibid., 180), Malthus direly predicts that

> in so short a period as within fifty years violence, oppression, falsehood, misery, every hateful vice, and every form of distress, which degrade and sadden the present state of society [will] have been generated by the most imperious circumstances, by laws inherent in the nature of man, and absolutely independent of all human regulations.
>
> (Ibid., 191)

This is typical Malthusian excess, but alas, Godwin's only direct argument against the principle of population is not much better. His faith in moral perfectibility is such that it leads him to conjecture that, by the time humanity is faced with an eventual limit of resources (in myriads of centuries), man's infinite rational and intellectual capacity to rise above bodily wants will result in the "extinction of the passions between the sexes." Malthus does not have too much trouble demonstrating that no such tendency has been observed in the past, nor is likely to arise in the future. But the long chapters he subsequently expends to rebut so weak a proposition turn into a contest between extremes, as futile as it is relentless, where in response to Godwin's extravagant claims, for the potential immortality of man among others, Malthus comes close to denying any real chance for human perfectibility, particularly in the social realm. It is in those chapters that he is exercised into some of the pronouncements that have most contributed to his reputation for conservative meanness, such as this:

> Were I to live a thousand years, and the laws of nature to remain the same, I would little fear, or rather little hope, a contradiction from experience, in asserting, that no possible sacrifices or exertions of the rich, in a country which had been long inhabited, could for any time place the lower classes of the community in a situation equal, with regard to circumstances, to the situation of the common people, about thirty years ago, in the northern States of America.
>
> (Ibid., 277)

Condorcet's challenge is more serious. To be sure the specific text Malthus has selected for criticism, the ultimate chapter of the *Esquisse* evoking the tenth "Period," the final, speculative stage of Condorcet's history of human progress, is not without its own excesses. Written in a feverish spirit, as enthusiastic about the future as the author's present circumstances were desperate, it is an unparalleled act of faith in humanity, no less idealistic than Godwin's. But Condorcet was a thinker to contend with, a good mathematician, a contributor to the *Encyclopédie*, for many years Secretary of the French Academy of Sciences, and as already mentioned, a younger friend of Turgot's and his first major biographer. Malthus himself recognizes this, as he concludes his two chapters largely devoted to the French *philosophe* by remarking:

> Mr. Condorcet's book may be considered, not only as a sketch of the opinions of a celebrated individual, but of many of the literary men in France, at the beginning of the revolution. As such, though merely a sketch, it seems worthy of attention.
>
> (Ibid., 172)

This attention, however, is almost entirely negative on the part of Malthus, and from the perspective of posterity at least, it does not honor him. The first of Condorcet's proposals he chooses to dismiss is a strikingly forward-looking system of life and old-age insurance for the laboring poor, based as its modern counterparts are on statistics, "the application of calculations to the probabilities of life" (ibid., 147). This was one of the earliest such plans to be seriously conceived, with the help and inspiration of an older pioneer in the field of socially applied mathematics, Richard Price, who was, not coincidentally, another of Malthus' nemeses.[15] Malthus rejects it on the same grounds that he chastises Godwin's collectivism. Under such a protection from the threat of hardship for all working people and their dependents, even for "the idle and the negligent," all the restraints to a population explosion would be sure to be removed, with the aggravating factor that "the labour necessary to procure subsistence for an extended population, will not be performed without the goad of necessity" (ibid., 149). This is the same logic, indeed the same principle of population, that in real life made Malthus an ardent opponent of the English Poor Laws – to be fair a very flawed system[16] – and from which Godwin could not offer a credible defense.

Only Condorcet does have a credible answer to the principle of population, albeit one that Malthus cannot hear. We have seen that Condorcet's belief in the benefits of scientific progress is almost infinite, and in this he is on the side of those among our contemporaries who remind us never to underestimate the potential of yet unforeseen technological discoveries. But he is also capable of imagining a time when all our ingenuity will not be sufficient to overcome limits. In a passage Malthus quotes at length, and approvingly, Condorcet evokes just such a possibility:

> But in this progress of industry and happiness, each generation will be called to more extended enjoyments, and in consequence, by the physical constitution of the human frame, to an increase in the number of individuals. Must not there arrive a period then, when these laws, equally necessary, shall counteract each other? When the increase of the number of men surpassing their means of subsistence, the necessary result must be, either a continual diminution of happiness and population, a movement truly retrograde, or at least, a kind of oscillation between good and evil? In societies arrived at this term, will not this oscillation be a constantly subsisting cause of periodical misery? Will it not mark the limit when all further amelioration will become impossible, and point out that term to the perfectibility of the human race, which it may reach in the course of ages, but can never pass?
>
> (Ibid., 151–152)

Condorcet's first answer to his own question is less to Malthus' liking, since for the latter, such a term is always imminent. For Condorcet, on the contrary:

> There is no person who does not see how very distant such a period is from us; but shall we ever arrive at it? It is equally impossible to pronounce for or against the future realization of an event, which cannot take place, but at an aera, when the human race will have attained improvements, of which we can at present form no conception.
>
> (Ibid., 152)

That is not the end, however, nor the more revealing part of their virtual dialogue. As Malthus then puts it:

> Mr. Condorcet, however, goes on to say, that should the period, which he conceives to be very distant, ever arrive, the human race, and the advocates for the perfectibility of man, need not be alarmed of it. He then proceeds to remove the difficulty in a manner, which I profess not to understand. Having observed, that the ridiculous prejudices of superstition, would by that time have ceased to throw over morals, a corrupt and degrading austerity [Condorcet's actual words: an austerity that corrupts and degrades morality instead of purifying and elevating it], he alludes, either to a promiscuous concubinage, which would prevent breeding, or to something else as unnatural.
>
> (Ibid., 153–154)

The allusion, which Malthus coyly professes not to understand, though he understands it perfectly (but does not quote it in his otherwise reasonably faithful translation), is quite clearly an endorsement of deliberate birth control:

> men will then know that, if they have obligations to [human] beings that do not yet exist; those do not consist in giving them existence, but happiness; that their object is the general well-being of the human species or of the society in which they live, of the family to which they are attached, and not the puerile idea of overloading the earth with useless and wretched beings.
>
> (Condorcet, 1795, 282)

For as is well known, Malthus is also adamantly opposed to any practice, aside from "restraint," meaning abstinence before marriage, that would interfere with the natural procreative process. The very words he uses to surmise Condorcet's suggestion, "promiscuity," "concubinage," or other "unnatural" (we presume sexually perverse) behavior, are typical of the way he always characterizes such practices, all of which fall for him under the category of "vice." Why promiscuous concubinage would be any less favorable to procreation than the marital bed is not at all clear, unless Malthus assumes that its practitioners are likely to be more canny about preventive methods.[17] But that is beside the principal point, which is his total rejection of the only practice that could humanely remove the pressure from the principle of population. And this not only on strictly moral and religious grounds (Malthus was a parson in the Anglican Church), but also because, as we now understand from his social anthropology, he is not altogether persuaded not only that this pressure could be removed, but that it should be.

For Malthus' legacy, this combination of casting a natural curse on man with a moral stricture on the means to escape from that curse is devastating, even if (or all the more because), for cultural reasons, it took much longer for the second part of that double-binding proposition to align with the Left–Right political polarity emphasized earlier.[18] But before returning to that legacy, we should not quite leave the Malthus/Condorcet controversy without noting another exchange that has important bearing for the general question we are pursuing. For in the second of those two chapters, the discussion rises for a moment much closer to the level of a theoretical debate on the notion of limits.

At question is the meaning of the word indefinite. Condorcet, like Godwin but from a more scientific perspective, speculated that with the progress of medicine the average lifespan of man could be prolonged with no assignable limit. His exact formulation is: "Certainly man will not become immortal, but may not the distance between the moment when he begins to live, and the common period when, naturally, without illness, without accident, he finds it difficult to continue existing, increase without ceasing?" (Condorcet, 1795, 294). Or a few lines above:

> Now would it be absurd to suppose that this improvement of the human species must be considered susceptible of an indefinite progress, [...] and that finally the duration of the average interval between birth and destruction may itself have no assignable term?
>
> (Ibid.)

A proposition Malthus summarizes fairly accurately as:

> that though man will not absolutely become immortal, yet that the duration between his birth, and natural death [what we call life expectancy], will increase without ceasing, will have no assignable term, and may properly be expressed by the word indefinite,
>
> (Malthus, 1798, 156)

the last word being Condorcet's own in the earlier expression "progrès indéfini." Of course, Malthus disagrees, remarking once again that no such tendency has been observed thus far, and that there is a natural limit to the human lifespan that will never increase significantly. There again he is wrong, but not as wrong as we have seen him in other predictions, depending on what one measures. Average life expectancy has roughly doubled in comparable countries since his own time, though it is extremely unlikely to keep increasing in such proportions, but on the other hand the longest lives ever achieved by any individuals have not increased greatly in historical times. This, however, is not the most interesting part of the discussion. At the point where we left him in his exposition, Condorcet felt it necessary to distinguish precisely between two meanings of the word indefinite. In Malthus' concise version: "He then defines this word to mean, either a constant approach to an unlimited extent, without ever reaching it; or, an increase in the immensity of ages to an extent greater than any assignable quantity" (ibid., 156–157). Although shorter than the original, this does not seem to betray Condorcet's meaning. In neither sense is the increase infinite, although in the first it is nearly so, the mathematical definition of infinity: "a law such as it continually approaches an unlimited extent, without ever being able to reach it." It is however in the second sense that Condorcet speaks of an indefinite increase of the human lifespan: "a law such as this same duration may acquire [...] a greater extent than any determined quantity whatsoever that would have been assigned as its limit." Greater than any given, previously assigned limit, not in any sense limitless. And Condorcet adds: "In this latter case, these increases really are indefinite in the most absolute sense, since there does not exist any boundary beyond which they must stop" (Condorcet, 1795, 294–295). Malthus nevertheless continues his intended refutation, citing various sensible examples from the natural world, but in a manner that seems to miss the theoretical point. "The sandy foundation on which the argument rests," he insists, "is, that because the limit of human life is undefined; because you cannot mark its precise term, and say so far exactly shall it go and no further; that therefore its extent may increase for ever, and be properly termed, indefinite or unlimited" (Malthus, 1798, 162–163). And a few pages later: "The error does not seem to lie, in supposing a small degree or improvement possible, but in not discriminating between a small improvement, the limit of which is undefined, and an improvement really unlimited" (ibid., 170). Malthus is so bent on this point, that he will reiterate it even more forcefully in a much later chapter on Godwin:

> And I cannot help again reminding the reader of a distinction, which, it appears to me, ought to be attended to in the present question; I mean, the essential difference there is, between an unlimited improvement, and an improvement the limit of which cannot be ascertained.
>
> (Ibid., 272)

But this distinction between indefinite and unlimited, the two terms Malthus keeps conflating in his refutation, is precisely the one Condorcet had sought to establish. The disagreement, at least at the theoretical level, seems to be based on a misunderstanding.[19]

If that is the case, it is a shame, because for once the two antagonists were in agreement on an analytical concept that is eminently applicable to the question of limits. Most limits we have to contend with do not present themselves as an immovable wall, but rather as a threshold around which it becomes increasingly more difficult, or costly, or destructive to progress, and for which the word indefinite is not a bad term, as long as we understand it as meaning something very different from infinite. That Malthus agreed with this distinction is all the more credible when we remember that, as an economist, he was one of the originators of another, related, concept, this time absolutely fundamental not only for an understanding of limits but for economics altogether; the law of diminishing returns, though this discovery had already been anticipated by Turgot and is more conventionally associated with the name of Ricardo. It is hardly an overstatement to say that before marginal analysis, there had not been a more important development in economic theory. Although the law does not apply to everything – an infinitely available resource whose increased use does not require proportionately greater effort could be immune to it – it does apply to almost all processes in which at least one of the factors is limited, or that have increasingly negative consequences, that is to say nearly all human interactions with nature. In fact, Malthus gives us numerous examples of thinking in terms of diminishing returns in the *Essay*, not only when speaking of limits to agricultural productivity (the problem for which the law was first formulated), but also with such amusing absurdities as trying to breed flowers so big that their stem could no longer support them, or sheep with legs so thin that that they would collapse under their own weight. He even gives us, in a chapter on Godwin, extrapolations to our ordinary physical and moral experience:

> Intemperance in every enjoyment defeats its own purpose. A walk in the finest day, through the most beautiful country, if pursued too far, ends in pain and fatigue. The most wholesome and invigorating food, eaten with unrestrained appetite, produces weakness, instead of strength. Even intellectual pleasures, though certainly less liable than others to satiety, pursued with too little intermission, debilitate the body, and impair the vigor of the mind.
>
> (Malthus, 1798, 212–213)

Had Malthus remained on this level in engaging with his progressive antagonists, or on that of his later, more sober economic work – for which, it is worth repeating, he is respected, and has even famously earned the lavish praise of Keynes for his early recognition of the possibility of crises caused by overproduction, or "glut"[20] – he may have served the cause of limits well. But as we know, it is not for these contributions that he is most remembered, except by specialists, but for his more apocalyptic pronouncements, which moreover have mostly turned out wrong, and often seem heartless. We have seen several, but one last iteration may be a useful reminder:

> Were a country never to be over-run by a people more advanced in arts, but left to its own natural progress in civilization; from the time that its produce might be considered as a unit, to the time it might be considered as a million, during the lapse of many hundred years, there would not be a single period, when the mass of the people could be said to be free from distress, either directly or indirectly, for want of food."[21]
>
> (Ibid., 138–139)

That such statements antagonized most believers in human progress, and in particular the Left-leaning thinkers and reformists most committed to improving the condition of the vast laboring class, is not difficult to understand. Overwhelmingly, with some exceptions and complications to be noticed later, the name of Malthus and the notion of unchangeable natural limits indelibly attached to it became anathema to the large and diverse socialist movements that were to grow in response to the industrial revolution, a justifiable rejection that has had immeasurable and lasting consequences.[22] This is not to suggest that without Malthus the history of limits, and in particular the position of the Left within that history, would have been radically different. Apart from the sheer improvability of such a thesis, there is on the contrary a strong sense that after the turning point evoked in the last chapter, the whole movement of progress in Western modernity, enabled by the scientific and then industrial revolutions that sustained it, has been an irresistible trend away from the traditional limits, natural and social, that were in the process of being overcome. Even the greatest thinkers ride such tides rather than create them. But at the very least Malthus caused some revealing turbulence within this current, and made some waves of his own, undoubtedly serving to inflect and perhaps amplify its drift toward the Right. In this he was famously helped by the public antagonism of a greater and even more consequential mind, with whom we will also have to contend in the next chapter; and this is of course Karl Marx. It never was a good idea to get on the wrong side of Marx's ire, but the torrents of contumely poured on Parson Malthus by the founder of Marxism are legendary. As a small sample:

> What characterizes Malthus is the *fundamental meanness* of his outlook; a meanness which only a parson could permit himself to display, a parson who looks upon human misery as the punishment from the Fall of men and stands in general need of "an earthly vale of tears," but who at the same time, out of consideration for the benefits accruing to him, finds it most advantageous, with the help of the dogma of predestination to "sweeten" the sojourn of the ruling classes in this vale of tears.
>
> (Meek, 1954, 118)

It is evident that between the two thinkers the chasm – philosophical, moral, social and political – is abysmal. Surely no less than the sentimental Godwin, Marx was one of those theorists "attributing almost all the vices and miseries that are seen in civil society to human institutions," and furthermore aiming to give this attribution a scientific backbone even more opposed to Malthus' theological naturalism. But it is worth remarking that Marx's contempt is not primarily based on economics. With Ricardo, whose economic analysis Malthus largely shared in spite of their public difference (Ricardo was also largely supportive of Malthus' thesis), and who was in many ways Marx's theoretical master, the latter, when he does, always disagrees respectfully, and often with praise. Not so with Malthus, for whom his scorn is almost visceral, deeply rooted, one senses, in class identification and cultural antipathy. But whatever the reasons – and we will revisit the more substantial of those later – this declared antagonism has had an enormous and well-documented influence. First, on the Marxist tradition and communist movements themselves, all the way to Lenin and the Russian revolution. And beyond this tradition, on the many, fellow travelers or more distant followers, who have over the years been sufficiently impressed with its theoretical analysis, and even among those who repudiate the theory, on the many more who share its demand for greater social justice and equality. For a very long time, and it may still be largely true to this day, any invocation of (Malthusian) limits was suspected by the Left of being part of the conspiracy of the (capitalist) rich, or the privileged West, against the wretched of the earth. Limits nearly became a Leftist taboo. This does not mean, it should be added, that a newer, greener environmental discourse of limits coming more recently from the Left cannot now be equally scorned by the liberal Right for its economic naivety, thus resulting in a virtual consensus against the recognition of our common finitude: *C'est la faute à Malthus*.

Neo-Malthusianism is a different story, but one that ironically also contributed for a long time to giving Malthus a bad name, on the Left almost as much as on the Right, although this time quite unjustifiably. For what became known as Neo-Malthusianism, the advocacy of contraception as a means for controlling the size of (principally poor) families, goes decidedly against Malthus' intentions and moral beliefs, and it was for the most part born on the left, even radical side of the socio-political spectrum.[23] It is always risky to draw generalizations about such a diverse movement that at different times and in different countries included social activists, hygienists, and feminists, whose aims ranged from prudent reformism to class warfare and sexual revolution,[24] but it would not be too wrong to see it occupying throughout the nineteenth century a marginal position similar to that of present day *altermondialistes*.[25] Of course it was shocking to conservatives, but because traditional sexual morality and gender conservatism was by no means exclusively on the Right, militant Neo-Malthusianism was also met with general and long-lasting hostility by most of the socialists themselves, Marxists and non-Marxists alike. To further push a risky comparison, Neo-Malthusian radicalism was to socialism then what anti-growth radicals are to the mainstream Left today. If this historical analogy turns out to have historical validity (and let us admit that it is a big *if*), it might be a source of considerable hope for the future of the *décroissance* movement. For as we can observe on a worldwide scale, the tides against Neo-Malthusianism have gradually but massively turned. After nearly a century and a half of generalized opposition to population control coming both from various breeds of natalist nationalisms and populist socialisms, and following the initial phase of access to rising prosperity that is normally favorable to population, the traditional ideal of having as many children as possible is increasingly being replaced, in one area of the world after another, by the "bourgeois" norm of smaller, more comfortable families, approaching the crucial replacement rate of two offspring per mother. Thanks to the pioneering work of those once marginalized advocates of family planning called Neo-Malthusians, and more decisively as a result of an irresistible combination of economic development, rising levels of education, and feminism, humanity has finally entered into what is known as the demographic transition. And that, by any estimation, has to be one of the most hopeful and momentous developments in the history of limits. Apart from religious fundamentalists of various persuasions, and a number of still very traditionalist societies (sometimes but not always in the same countries), the whole world is becoming Neo-Malthusian.

Notes

1 Hence the Malthusian League, founded in 1878, and its journal the *Malthusian*, published from 1879 to 1922.

2 As Robert Mayhew certainly is in his well-regarded *Malthus: Life and Legacy of an Untimely Prophet*, much of which is devoted to a careful examination of the various currents of hostility to Malthus – which he also calls "moments" – as well as to a generous defense of Malthus' position (Mayhew, 2014).

3 See for example Charles Emil Stangeland: *Pre-Malthusian Doctrines of Population* (Stangeland, 1966) and Joseph J. Spengler: *French Predecessors of Malthus* (Spengler, 1942).

4 In case there is any doubt in the reader's mind, I am here deliberately echoing Michel Foucault's famous evocation of Don Quixote at the beginning of Chapter 3 of *The Order of Things* (Foucault, 1973, 46–50).

5 W. W. Rostow, who is unquestionably knowledgeable and always fair minded, concludes the section on Malthus in *Theorists of Economic Growth* with this surprising judgment:

> The irony of all this is that of the four great classical figures – Hume, Smith, Malthus, and Ricardo – I would guess that Malthus felt most deeply and personally – least abstractly – the reality of human misery among the poor and found the greatest difficulty in reconciling that reality with belief in a just God.
>
> (Rostow, 1990, 72)

6 As Kenneth Smith writes in *The Malthusian Controversy*:

> This sounds impressive [...] but reflection reveals that the appearance is deceptive. Since it is impossible to imagine the situation of the third term of the series, it is needless to go further [...] Therefore, since the third term can never come into existence, all the subsequent terms are false [...] If food could only increase in arithmetical ratio, so could population.
>
> (Smith, 1951, 233–234)

7 See Maurice Godelier: "Les sources ethnographiques de la pensée de Malthus" (Fauve-Chamoux, 1984, 127–145). According to Robert Mayhew's count, whereas the 1798 edition of the *Essay* cited only 11 other works, that number went up to 125 in the second edition of 1803, and by the sixth and last revision, in 1826, it had grown to 162 (Mayhew, 2014, 111).

8 Robert Mayhew insists that this was not Malthus' intention, in spite of the apparent harshness of the first *Essay*, and that by the time of the later editions and of the *Principles* (1820) his position has softened considerably with regards to the possibility of maximizing the "preventive" checks (through the exercise of "prudence,") and minimizing the "positive" checks, famine, war, and pestilence: "Malthus's enlightened history of social progress is itself the narrative of the changing power of these two checks, of the growth of the preventive checks and the diminution of the immediate power of positive checks" (Mayhew, 2014, 71).

9 On this point too Malthus' position seems to have evolved considerably. According to Robert Mayhew: "... the limits Malthus attended to in his later thought were not absolute but instead those created by social circumstances." And he quotes a passage from a later edition of the *Essay* that we could well have found almost word for word in Adam Smith:

> no country has yet reached, or probably will ever reach, its highest acme of produce, it appears always as if the want of industry, or the ill direction of that industry, was the actual limit to a further increase of produce and population, and not the absolute refusal of nature to yield any more.
>
> (Mayhew, 2014, 121)

10 All excerpts from Condorcet's *Esquisse* except for those quoted by Malthus himself are in my slightly awkward but faithful translation.

11 See for example Michelle Perrot: "Malthusianisme et socialisme" (Fauve-Chamoux, 1984, 357–369), and Annie Vinokur: "Malthusian Ideology and The Crises of the Welfare State" (Turner, 1986, 171–186).

12 Here we can only acknowledge Robert Mayhew's strongly dissenting view:

> In its own age and since, [the *Essay*] has been misprisioned as conservative due to the antirevolutionary ambitions with which it was penned in 1798, but in fact the *Essay* reveals the mind of a man who had been educated in ways scornful of any simplistic binary between "conservatism" and "radicalism," as the closing religious chapters demonstrate.
>
> (Mayhew, 2014, 74)

13 To be fair, by the time he published Book II of his *Principles of Political Economy* (1820), Malthus seemed to have softened his position considerably, and granted even the lower classes the hope of escaping from the curse of misery by exercising restraint, the possibility of choosing to "act as beings who look before and after, and who consequently cannot acquiesce patiently to the thought of depriving themselves and their children of the means of being respectable, virtuous, and happy." And among the social factors that would favor this moral improvement, he advocates "civil and political liberty, and education" (Malthus, 1820, 225–227). This brings him much closer to Condorcet's position, but over 20 years after the first publication of the *Essay*, in a scholarly treatise that many fewer people could read. One is tempted to think that the damage was done.

14 As his father was an admirer of Jean-Jacques Rousseau, Malthus would have been well aware of the progressive thinkers of his time, notably Richard Price (1723–1791), whom he had clearly read. Since he took William Godwin (1756–1836) seriously enough to refute his ideas, he must also at least have known of Mary Wollstonecraft (1759–1797), the most famous feminist of her time. And among his contemporaries just a few years younger, was Francis Place (1771–1854), the first open propagandist of birth control.

15 Richard Price, Robert Mayhew informs us,

> had been corresponding with Condorcet for a number of years by the time of the momentous events of 1789 about shared interests in what we might, as a shorthand, call "radical demography," the quintessentially enlightened idea that the application of mathematical reasoning to the amelioration of social problems was the key to unlocking the door to utopian progress.
>
> (Mayhew, 2014, 36)

16 The "radical defect of all systems of this kind," Malthus objects, is that they tend "to increase population without increasing the means for its support, and thus to depress the condition of those that are not supported by parishes, and, consequently, to create more poor." Or more succinctly: "They may be said therefore in some measure to create the poor which they maintain" (Malthus, 1798, 94–95, 83).

17 John Bellamy Foster suggests: "... one may surmise that Malthus held to the characteristic eighteenth-century belief – explicitly stated by Godwin – that "the promiscuous intercourse of the sexes" itself constituted a preventive check on population (Foster, 2000, 97). John Avery has remarked with regard to Condorcet that: "Probably his belief was based on observation, since what are today considered to be minor venereal diseases would often produce sterility in Condorcet's time" (Avery, 1997, 11).

18 The birth control movement really became popularized and associated with social if not socialist advocacy only after the Charles Bradlaugh and Annie Besant trial of 1877–1879. Francis Place, however, had initiated the first mass education campaign

in 1823, three years before the last edition of Malthus' essay, and Jeremy Bentham was already recommending the use of contraceptive methods for the reduction of poverty as early as 1797, a year before the first edition (Himes, 1963, 194).

19 See Hervé Le Bras: "Malthus et les deux mortalités" (Fauve-Chamoux, 1984, 54).

20 He is the first economist to have identified and made extensive use of the "principle of effective demand." For Keynes and Malthus see Robert Mayhew's *Malthus* (Mayhew, 2014, 167–173).

21 And I have refrained from quoting what is probably Malthus' most infamous passage, the parable on "Nature's mighty feast" added in the 1803 edition of the *Essay* (Malthus, 1803, 249). Whatever his original intention, Malthus himself removed it from all subsequent revisions.

22 In addition to the articles cited in note 11 (this chapter), see in the same conference proceedings Yves Charbit: "Le Destin de l'oeuvre: Histoire et Idéologie" (Fauve Chamoux, 1984, 49–59).

23 See in particular William Petersen: *Malthus, Founder of Modern Demography* (Petersen, 1979, Chapter 9 and following) and Kenneth Smith: *The Malthusian Controversy* (Smith, 1951, Chapter 4).

24 See again Michelle Perrot: "Malthusianisme et socialisme" (Fauve-Chamoux, 1984, 362 and following).

25 Except for the fact that, marginality aside, it was much more dangerous to be an open advocate of birth control. Not only were its propagandists vilified by all manner of public opinion, but they were also put on trial by official courts of law on both sides of the Atlantic, not merely ignored as the *altermondialistes* mostly are.

References

Avery, John (1997) *Progress, Poverty and Population: Re-Reading Condorcet, Godwin and Malthus*. London: Frank Cass.

Blum, Carol (2002) *Strength in Numbers: Population, Reproduction, and Power in Eighteenth-Century France*. Baltimore and London: The Johns Hopkins University Press, 2002.

Condorcet, Marie-Jean-Antoine-Nicolas Caritat, Marquis de (1795) *Esquisse d'un tableau historique des progrès de l'esprit humain*. Paris: Flammarion (this edition 1988).

Ehrlich, Paul (1968) *The Population Bomb: Population Control, or Race to Oblivion*. New York: Buccaneer Books.

Fauve-Chamoux, Antoinette (ed.) (1984) *Malthus hier et aujourd'hui: Congrès international de démographie historique CNRS, mai 1980*. Paris: Éditions du Centre de la Recherche Scientifique.

Foster, John Bellamy (2000) *Marx's Ecology: Materialism and Nature*. New York: Monthly Review Press.

Foucault, Michel (1973) *The Order of Things: An Archaeology of The Human Sciences*. New York: Vintage Books.

Himes, Norman E. (1963) *Medical History of Contraception*. New York: Gamut Press, Inc.

Malthus, Thomas Robert (1798) *An Essay on the Principle of Population*. Amherst, NY: Prometheus Books (this edition 1998).

Malthus, Thomas Robert (1803) *An Essay on the Principle of Population*. Cambridge: Cambridge University Press (this edition 1992).

Malthus, Thomas Robert (1820) *Principles of Political Economy*. New York: Augustus Kelley (this edition 1951).

Mayhew, Robert J. (2014) *Malthus: The Life and Legacy of an Untimely Prophet*. Cambridge, MA and London: Cambridge University Press.

Meadows, Donella H., Dennis L. Meadows, Jørgen Randers and William W. Behrens (1972) *The Limits to Growth: A Report for the Club of Rome's Project on the Predicament of Mankind*. New York: Universe Books.

Meek, Ronald L. (ed.) (1954) *Marx and Engels on Malthus*. New York: International Publishers.

Petersen, William (1979) *Malthus, Founder of Modern Demography*. New Brunswick and London: Transaction Publishers.

Ricardo, David (1951) *Notes on Malthus's Principles of Political Economy. Works and Correspondence of David Ricardo*, vol. II. Cambridge, UK: Cambridge University Press.

Rostow, W. W. (1990) *Theorists of Economic Growth from David Hume to the Present: With a Perspective on the Next Century*. New York and Oxford: Oxford University Press.

Rothschild, Emma (2001) *Economic Sentiments: Adam Smith, Condorcet, and The Enlightenment*. Cambridge, MA, and London: Harvard University Press.

Smith, Kenneth (1951) *The Malthusian Controversy*. London: Routledge and Kegan Paul.

Spengler, Joseph J. (1942) *French Predecessors of Malthus*. Durham, NC: Duke University Press.

Stangeland, Charles Emil (1966) *Pre-Malthusian Doctrines of Population: A Study in the History of Economic Theory*. New York: Augustus M. Kelley.

Tobias, Michael (1994) *World War III: Population and the Biosphere at the End of the Millennium*. Santa Fe, New Mexico: Bear and Company.

Turner, Michael (ed.) (1986) *Malthus and His Time*. Houndmills, Basingstoke and London: Macmillan.

4 Final stages

In the nineteenth century, only one major economist, and a most classical economist too, embraced the end of growth, the voluntary acceptance of limits, as a positive goal for humanity,[1] indeed, if such a possibility can still be conceivable after our last chapter, as a beneficent version of Malthusianism. John Stuart Mill's "The Stationary State," a very short chapter deep into the fourth book of his *Principles of Economy*, sounds so extraordinary for its time, coming as it does from the liberal tradition, but also so surprisingly pertinent from our new perspective on limits, that it is well worth rereading with some attention after a period of relative neglect, though it never fell into complete oblivion. For it would be wrong to suggest that either the author, or the notion of stationary state itself, is in any way obscure. Mill was very famous and respected in his day, not only for his economics but also for his work on logic and his social thought, as well as his outstanding biography and, not coincidentally for his interpretation of Malthus, his championing of the rights of women. In fact, he is probably better remembered nowadays for these social ideas, and as a humanist, than for his contribution to mainstream economic thought, extensive though it was.

But the fate of Mill's political economy may not just have to do with the fact that by the time he wrote the *Principles*, his economics had become fairly traditional, not yet neo-classical, but rather late, harking back as they do to the already well-worn liberal line opened by Adam Smith, or that his name cannot as readily be identified with a distinct theoretical principle – besides that of the stationary state – as are those of a Ricardo, a Say, and even a Malthus. Thinking again in terms of the political polarization emphasized in the previous chapter, it is striking that in the middle of a century that was becoming increasingly polarized, Mill cannot easily be positioned on either side of the Left–Right divide. On the one hand he remained a faithful exponent of a moderate form of market capitalism, apparently confident in the objective soundness of its principles and in its economic benefits. On the other hand, he clearly believed, and fervently hoped, that this system would gradually evolve into a form of associative, or cooperative, but non-revolutionary socialism that had many other advocates in the second half of the nineteenth century. This could not sit well with the more radical partisans in either camp, nor greatly serve Mill's place in a history that tends to remember radicalism better than moderation. Marx, in particular, who had axes to grind with both liberalism and any brand of socialism not his own, was quick to dismiss him as another bourgeois economist, not indeed a nasty one like Malthus, but rather a sentimental one, which could be worse.

To understand the relative oblivion of the stationary state, however, calls for a much broader perspective than the merely political and polemical, one that will not take us away from Marx, since it leads us on the contrary to engage with the more substantial aspects of his thought, but rather requires that his own incomparably more influential vision of the final stage be replaced at the deeper level where the European conception of human progress reaches a near consensus in the nineteenth century, beneath all the very real differences as to where this progress would take us. In this perspective, the Hegelian philosophy of history, itself a stage theory that provides the intellectual framework for Marx's dialectic materialism and less explicitly for many other grand narratives, can be seen to be only the more noble, higher-end expression of the shared epistemological matrix of stadial progression. It would hardly be an exaggeration to say that virtually every major or minor socially scientific construct of that great era, from Comte to Bachofen and Morgan[2] and of course to Marx and Engels (Darwin, despite the common evolutionary aspect, is a very important departure from such stadial teleology) takes the form of a stage theory culminating in an inevitably superior end. The number of stages may vary greatly from the classic four of the Scottish Enlightenment model, and so can the emphasis on the original state and the nature of the progressive principle, more or less spiritual, organic or material, but what remains at the core is the combination of teleology with discontinuity, the latter providing the means to the end that ultimately transcends it. Hence a particularly fraught version of eschatology, where the final stage must perilously embody the very progressiveness it also ends. For such a model of progress to be consistent, the end of history must be open-ended. Which means limitless.

As we saw much earlier, this limitlessness of the last stage is the defining thrust of the liberal economic tradition, which on the contrary naturally sees the existence of limits, the possibility of an end to growth, as a challenge to be forever postponed – which is why John Stuart Mill's welcome of the stationary state from within that tradition, so early and so much against its grain, deserves to be given renewed prominence. But for the projectors of a more perfect state for humanity coming from the Left, particularly for those associated with the utopian tradition, and there were many in the nineteenth century, the reconciliation of open-endedness with finality, indeed with perfection, is more of a problem. It goes against their grain. Yet it is striking that even the most decidedly utopian among them, in whose schemes the whole theory is often reduced to a final stage, do their best to remain in a progressive mode. Unlike their forebears of the seventeenth and eighteenth centuries, for whom utopia was synonymous with closure, with finite numbers and regulated levels of (frugal) consumption, aiming in its perpetual repetition of the same ideality at a stasis of desire, nineteenth century cities down the hill wanted least of all to be thought of as dead-ends, affirming the opposite ideal of a social harmony based on the fullest expression of human aspirations. Fourier's Phalanstère,[3] with all its minutely calculated classifications designed for the free-play of the passions, may be the most extreme illustration of the tension between the utopian proclivity for a finite order and the progressive imperative of an infinite libidinal economy. Whether this tension can be resolved otherwise than in a utopian mode, whether it can be resolved at all, is the great question faced by any dreamer of a perfect state who would deny neither the infinity of human desire nor the finiteness of the world.

Karl Marx would not have appreciated being put in the company of utopists, and for good reasons. Many of the differences by which he strenuously sought to distinguish himself from idealistic schemers are very real and profound. For one thing he, unlike most of them, did pay a lot of attention to earlier stages, most prominently to the one great stage he spent his whole life analyzing with such critical acumen, capitalism, as well as to the dialectical process that would famously lead it to self-destruct into the last and final stage of communism. Yet for all these differences, and for all his assurance of scientific objectivity, Marx could not escape any more than his idealistic competitors from the common problematic of progressivism and finality, though he undoubtedly thought it would eventually be resolved with the inevitable advent of his own vision of the final stage. Given the incomparable influence of that prophetic vision, so great that for more than a century it all but eclipsed Mills' stationary state along with every other alternative outside of liberalism, Marx must be acknowledged not just as the fierce antagonist and critic we saw in the previous chapter, but for the role he positively played in the history of limits, even if this role, especially as relayed by "vulgar" and practical Marxism, has so far amounted largely to a denial of finitude. More recent interpretations in the light of our new environmental consciousness suggest that his contribution is a more complex one, and that his role may not yet be over.[4]

For the fundamental materialist premise of Marxism, it must be remembered, the materialism it inherited from its predecessors in the Four Stage Theory, is not initially inauspicious for environmentalism. What it means is that human beings are first and foremost defined by their interaction with the natural world, by the concrete ways in which they earn their subsistence and make their living from their environment. The history of humanity is basically nothing other than the gradual unfolding of this economic-ecological engagement with nature. It is this material foundation that for Marxism ultimately determines culture and social consciousness, so that human beings cannot be conceived outside of their interrelatedness with nature. From his earliest philosophical pronouncements, Marx insisted on the organic unity of man and nature:

> Man lives on nature – means that nature is his *body*, with which he must remain in continuous intercourse if he is not to die. That man's physical and spiritual life is linked to nature means simply that nature is linked to itself, for man is a part of nature.
>
> (Tucker, 1978, 75)

To describe this essential connection between human beings and nature, Marx and Engels would later adopt the more scientific, indeed biological term of "metabolic relation (*Stoffwechsel*)."

Now it is true that our contemporary natural fundamentalists and deep ecologists would not find a philosophical ally in such materialism. If nature must in some sense exist in itself, there is no doubt that for Marxism, no less than in the Biblical tradition, it is there essentially *for us*, to be defined and shaped by the way we use it. Indeed, the very measure of progress in the Marxian stage theory is the extent to which humanity came to gradually exert an increasingly productive mastery over the natural environment. If Marx, and especially Engels in the most elaborate exposition of their stadial progression in *The Origin of The Family*, express unusual respect and even admiration for the egalitarianism and proto-communism of society at the first stage of "savagery," they also make it clear that it is an inferior stage, precisely because in this condition human beings are still subjected to nature, "dominated by it as a blind force," merely living off what it spontaneously produces without acting upon it. It is only with "Barbarism," whose lower tier in this nomenclature corresponds to the classic stage of pastoralism, and even more with its middle and superior stages, the emergence of agriculture and then metallurgy, with its attending division of labor into specialized crafts, that humanity begins to assume its proper role by gradually liberating itself from a passive dependence on the natural world and extending its dominion over it. Eventually, Marx proclaims, "Freedom in this field can only consist in socialized man, the associated producers rationally regulating their relation with nature, bringing it under their common control" (Elster, 1985, 85). Human progress in this light is an unabashed appropriation of the productivity of nature. No Gaia hypothesis for Marx.

But at the same time there is nothing in this initial vision that prevents a sense of limits, nor the possibility of imagining, and condemning, an exploitative or destructive relation to the natural world. In fact, Marx recognizes quite sensibly that "the productiveness of labour is fettered by physical conditions. These are all referable to the constitution of man himself (race, &c.), and to surrounding nature." He further makes among the latter, properly environmental factors a distinction that he perceptively identifies as a crucial determinant of stadial progression:

> The external physical conditions fall into two great economic classes, (1) Natural wealth in means of subsistence, i.e., a fruitful soil, waters teeming with fish, &c., and (2), natural wealth in the instruments of labour, such as waterfalls, navigable rivers, wood, metal, coal, &c. At the dawn of civilization, it is the first class that turns the scale; at the higher stage of development, it is the second.
>
> (Marx, 1867, I, 513)

The ultimate item listed, coal, and the tantalizing "&c." that follows, even suggest that Marx understood better than most of his contemporaries the looming importance of energy.[5] Indeed this impression of an unusual degree of prescience on the part of Marx is confirmed by an astonishing passage in a letter written to him by Engels, in which we find a clear recognition of the overexploitation (in 1882) of what we now call fossil energy. Human activity, Engels observes, "is not only a stabilizer of the *present* but also, and to a far greater extent, a squanderer of *past*, solar heat." And he adds in a generous acknowledgment of his friend's superior understanding of these matters: "As to what we have done in the way of squandering our reserves of energy, our coal, ore, forests, etc., you are better informed than I am" (Marx and Engels, 1996, vol. 46, 411). Now it could be that apart from this last point, we would find nothing very remarkable about such statements of natural constraints, except for the fact that it is Marx who is clearly making them – even if we are careful to notice that "fetters" are not quite the same as limits.

And if we were still doubtful about the possibility of an environmental Marx, and even more of his imagining what we now describe as an overshooting of ecological limits, the following passage from another letter, this time from himself to Engels (March 25, 1868), might be equally arresting. Commenting on a pioneering book by C. [Karl Nikolas] Fraas, "a Darwinist before Darwin," on the evolution of climate and plants,[6] Marx reports approvingly the claim that

> with cultivation – depending on its degree – the "moisture" so beloved by the peasants gets lost (hence also the plants migrate from south to north), and finally steppe formation occurs. The first effect of cultivation is useful, but finally devastating through deforestation etc. [...] The conclusion is that cultivation – when it proceeds in natural growth and is not consciously controlled [...] leaves deserts behind it, Persia, Mesopotamia, etc., Greece.
>
> (Marx and Engels, 1996, vol. 42, 559)

Long before Joseph Tainter, Jared Diamond and contemporary eco-prehistorians, Marx had therefore identified cases of "collapse" due to overexploitation of natural resources. What is more, he does not see such cases as mere learning mistakes on the part of past civilizations, but as a fundamental betrayal of our common custodial responsibility, if not to nature itself, at least to the future share of humanity. This is made clear by two other remarkable passages, in this instance not from his private correspondence, but for all to read in the last volume of his *magnum opus*. In agriculture as it has historically been practiced, i.e., based on private ownership, Marx writes:

> exploitation and squandering of the vitality of the soil [...] takes the place of conscious rational cultivation of the soil as eternal communal property, an inalienable condition for the existence and reproduction of the chain of successive generations of the human race.
>
> (Marx, 1867, III, 812)

And about 100 pages later:

> Even an entire society, a nation, or all simultaneously existing societies taken together, are not the owners of the earth. They are simply its possessors, its beneficiaries, and have to bequeath it in an improved state to succeeding generations, as *boni patres familias*.
>
> (Ibid., 911)

These are strong statements, and they provide unequivocal evidence of an environmental consciousness in Marx, and of his recognition that there are natural limits humanity can only transgress at its own peril.

But they are not universal statements. In each case we should also give all possible weight to the crucial qualifications that condemn agriculture to exploitation and devastation: when it is not "consciously controlled," and again when it "takes the place of conscious rational cultivation," that is to say under the regime of private property that substitutes individual profit for the communal care of the land. A reminder that there is no natural relation to nature in Marx, only contingent possibilities thoroughly mediated by social modes of production. In fact, it has often been remarked that nature in this view is but a mirror reflecting the human world itself, our relation to nature becoming the image of the power and labor relations prevailing among ourselves. The grand narrative of stage progression inspired by Marx and Engel's reading of Morgan's *Ancient Society* contains powerful examples of such parallels, even though the modeling sometimes seems to operate in the other direction. Thus it is the domestication of animals and the later harnessing of their power for heavy agriculture that gives rise to human slavery, not only by creating a use for the labor of the conquered that did not exist at the hunting and gathering stage, but also by providing the very model for enslavement itself. Engels, as decidedly a conscious feminist as we will find Mill, even extends the homology to the power relations between the sexes. It is also with the human domination over animals and the substitution of deep plowing for horticulture that women were displaced from their formerly more equal (if different) role as producers and relegated to a subaltern, dominated position in the family. Husbandry means the subjugation by men of both nature and women.

It stands to reason that from this perspective capitalism, the socially exploitative regime par excellence, would be found to be equally exploitative of natural resources. We can therefore say with some assurance, and the few explicit statements we have seen confirm it, that for Marxism authentic environmentalism is incompatible with the capitalistic organization of society – a warning that should perhaps give pause even to those who do not agree with the Marxist solution. But does the converse proposition follow? Would a society freed from human exploitation and the profit motive also refrain from "exploiting" nature? Here we have to be careful to avoid a conceptual homology that does not make sense from the point of view of Marx's materialist humanism, for "nature" is not a moral entity with its own rights and protections; it is precisely there to be exploited by man, even if for the good of human posterity such exploitation should be "consciously controlled" and "rational," terms we may now want to translate as "sustainable." Which brings us back to the question of limits, a question that within the Marxist conception of nature may well remain unanswerable, if it can be posed at all.

One of the most important differences between the Marxian theory of stadial progression and its classical antecedents, more profound indeed than the immediately noticeable nomenclature of the stages themselves, is that, unlike Adam Smith's classic version in particular, it is not premised on a fundamental lack in the natural condition of human beings, an original scarcity from which they must escape in order to survive, but on the contrary moved forward by our productive capacity in a plentiful world. Even at the first stage, when hunters and gatherers enjoy only a minimal level of subsistence, living day to day without surplus from the spontaneous productions of nature because they have not yet learned to become themselves producers, they are in a state of abundance, untapped yet potentially limitless. Lack in the Marxian reading of the world is neither originary nor natural; it is on the contrary the result of the historical process whereby resources and productive forces are at once greatly multiplied and appropriated by the few at the expense of the many. The same economic progress that has so dramatically increased man's overall productivity is also what created inequality among mankind and scarcity for the vast majority of the exploited.[7] The limitedness so far experienced by human beings is therefore not inherent in the world, but a function of the modes of production through which we have interacted amongst ourselves and with the world. Scarcity in this view is (almost) entirely relative to the exploitative organization of society. It follows that once human beings freed themselves from social oppression and inequality while retaining the productive advances made possible by industrial capitalism, they would also free themselves from limits, their productive power becoming "immeasurable." They would live, Marx and Engels seem to conclude in the *Critique of the Gotha Program*, in a state of "overabundance," a state of near limitlessness.

Or such would be, if we are to believe the mainstream of the Marxist tradition, the final stage of human development. And indeed, it is difficult to imagine a world in which every person is able to realize without social limitations his or her full spiritual, creative, and material potential, as was clearly Marx's profoundly humanistic ideal, without also believing in humanity's potential to produce commensurately unlimited means. For each to receive "according to his needs" from each "according to his ability," our collective ability must be at least as great as our multitudinous needs, and so must presumably be the ability of nature itself. Even if in this condition freed from alienated desire and the fetishization of consumption human beings became wiser and more rational about their needs, this would seem to mean that reaching the final stage of communism is not only incompatible with the recognition of limits, but based on their outright denial, or else on some of sort transcendence.

Is this a fair interpretation of what Marx had in mind, and how compatible is it with the evidence of environmental consciousness we saw earlier, however qualified by social relativism? Unfortunately, Marx, so prolific about earlier stages, wrote very little about the concrete workings of communism. Unlike the utopians from whom he wanted to mark his theoretical distance, he thought it futile to try to anticipate in detail a stage whose advent was historically inevitable and whose exact content was for future generations to determine. While this is consistent with his own principles, and certainly saved him from the embarrassing fate of most utopias, it may also mean that on the question of limits, his legacy remains equally open to the "Promethean" interpretation of orthodox communism as to that of the newer, greener Marxism.[8]

There is however, another theme running consistently and insistently throughout the entire span of Marx's writings that may give us a clearer sense, although no less challenging, of his environmental vision for the final stage. This recurring idea, that of a "metabolic rift" at the heart of modernity, brings us back to the materialist necessity of an organic link between human beings and nature, and is in fact closely connected with one of the most central concepts of the Marxist view of modernity, that of alienation. It is in a first sense well known that for Marx, under the capitalist mode of production, workers are alienated from the process and product of the labor they provide for others through means no longer their own or under their control. In a second, more fundamental sense, they are alienated from the whole productive and creative process by the division of labor itself, whereby production is fragmented in so many disconnected and meaningless operations. But there is a third, no less profound sense in which Marx insists on the alienation of modern humans from the cycle of material production on which they all rely, and that is the separation between urban and rural life and activities, or between "town and country," which Marx in fact identifies as the "greatest division of material and mental labour." From the *Economic and Philosophical Manuscripts* of 1848 to his last writings, and in prominent passages of the most canonical texts of the Marxian opus, Marx (and Engels), never cease to denounce, first of all, the estranging, divisive, and stultifying effects of the separation between agricultural and city life on both rural and urban populations. Thus in *The German Ideology*:

> The antagonism between town and country [...] is the most crass expression of the subjection of the individual under the division of labour, under a definite activity forced upon him – a subjection which makes one man into a restricted town-animal, the other into a restricted country-animal, and daily creates anew the conflict between their interests.
>
> (Tucker, 1978, 45)

In the *Manifesto of the Communist Party*, they infamously wrote that by subjecting the country to the rule of the towns, and creating enormous cities, the bourgeoisie had "rescued a considerable part of the population from the idiocy of rural life" (Marx and Engels, 1996, 488), an often misunderstood and admittedly unfortunate sentence that can give the impression that only the "country animals" left behind suffer from the separation between town and country.

Numerous other passages make it clear however that the rescued "town animals" are equally negatively affected by their unnatural concentration in cities. Not only because of the appallingly insalubrious and crowded conditions in which the great majority of the new industrial masses were condemned to live in the working-class neighborhoods observed by Engels and Marx, but also because even the more fortunate among the city dwellers found themselves severed from the spiritual wellspring of the country, and nostalgically longing for it, as we can see to this day in the enduring attachment of the upper and even middle urban classes to the compensatory pleasures of a country home.

This is the human cost of the separation between town and country. But Marx insists at least as much on the ecological consequences of this estrangement, on the "metabolic rift" itself. By concentrating the population in urban centers, capitalist production "disturbs the metabolic interaction between man and the earth." Most concretely, "it prevents the return to the soil of its constituent elements consumed by man in the form of food and clothing; hence it hinders the operation of the eternal natural condition for the lasting fertility of the soil" (Marx, 1867, I, 637). When reiterating this same principle in *Capital* III, Marx will identify explicitly by name the principal scientific source of his and Engel's awareness of the ecological damage of modern agriculture, the German founder of soil chemistry Julius Liebig:

> an ever growing industrial population crammed together in large towns [...] produces conditions that provoke an irreparable rift in the interdependent process of social metabolism, a metabolism prescribed by the laws of life itself. The result of this is a squandering of the vitality of the soil, which is carried by trade far beyond the bounds of a single country (Liebig).
>
> (Ibid., III, 949)

From our post twentieth-century perspective, these passages may strike us equally as stunningly contemporary and almost pre-modern in their conservatism. On the one hand this is unmistakably the language of ecological consciousness and sustainability, ahead of its time by more than 100 years. On the other hand Marx, still thinking in terms of traditional agricultural recycling (he specifically talks about returning to the soil "the excrement produced by man's natural metabolism,") does not seem to anticipate the way in which the problem he rightly identifies would be addressed by the so-called "green revolution," the massive use of fossil fertilizers. But that is precisely what makes his thinking, and his own solution, so radically green.

For Marx does not aim to remedy the problems caused by the metabolic rift in modernity, but to put an end to it. The socio-environmental ambition for the final stage, as it is unequivocally announced in the programmatic section of the *Manifesto*, is nothing less than "the gradual abolition of the distinction ['contradiction' in the 1848 edition] between town and country, by a more equitable distribution of the population over the country" (Tucker, 1978, 505). And this can only be achieved by the "combination of agriculture with manufacturing industries," or as Engels put it in *The Housing Question*, by reestablishing an "intimate connection between industrial and agricultural production" and "as uniform a distribution as possible of the population over the whole country" (Engels, 1872, 92). That Marx and Engels absolutely mean this to be the outcome of the historical process of modernity, there cannot be any doubt. And once we recover from the audacity of such a proposition (putting aside for a moment the question of its plausibility), we realize that it is the only concrete resolution consistent with the conflict Marx posited from the beginning between the humanistic ideal of unity with nature and the alienating effects of capitalism (we might say of modernity itself), and with his vision of a final stage that "as fully developed naturalism, equals humanism, and as fully developed humanism, equals naturalism," achieves "the *genuine* resolution of the conflict between man and nature, and between man and man" (Marx, 1974, 348). And how else, we might half-facetiously ask, could I possibly "hunt in the morning, fish in the afternoon, rear cattle in the evening" and "criticize after dinner [...] without ever becoming hunter, fisherman, shepherd or critic" (Tucker, 1978, 160), as Marx famously wrote in the *German Ideology*, if the division between natural, rural, and city life were not also abolished? Just as inevitably as capitalist production has created a rift between town and country by tearing asunder "the old bond of union which held together agriculture and manufacture in their infancy" (Marx and Engels, 1996, 506), it has made it imperative for communism to repair the rift in what is clearly seen by Marx and Engels as a dialectical process, a synthesis that has become a social as well as an ecological necessity. In the equally programmatic *Anti-Dühring*, they proclaim:

> Abolition of the antithesis between town and country is not merely possible. It has become a direct necessity of industrial production itself, just as it has become a necessity of agricultural production and besides, of public health. The present poisoning of the air, water and land can be put an end to only by the fusion of town and country...
>
> (Marx and Engels, 1878, 361–362)

What would such a synthesis have looked like (a question, it seems, that we can only imagine posing in the past conditional, so unlikely has its eventual achievement become)? John Bellamy Foster, one of the most persuasive advocates for an ecological Marxism, maintains that we can find a good approximation of it in the conceptions of William Morris, himself an avowed convert to Marxism. Such a startling suggestion has at least the merit of making us rethink profoundly the way we have come to picture a fully realized communist society. But whatever its verisimilitude, we should resist the temptation of picturing a craftsman's paradise. In Marx and Engel's time, England had become the first country in the world where urban population exceeded the numbers in rural areas, and that in an overall territory already fairly densely populated. If one were to achieve (even remaining at mid-nineteenth century numbers) "as uniform a distribution as possible of the population over the whole country," the result would be the end of any semblance of rurality as we normally conceive it. And at any rate it is clear that Marx never envisaged anything like an Arcadian return, much less an extension to what he calls "the irrational, old fashioned methods of agriculture," but their replacement by "scientific ones." If communism was to overcome the divisiveness and destructivity of capitalism, it was precisely by taking advantage of the latter's unprecedented productivity and scientific achievements. In the crucial passage of *Capital* quoted earlier, where we saw Marx denouncing capitalist production for "tearing asunder" the former union between agriculture and manufacture, he nevertheless continued:

> But at the same time it creates the material conditions for a higher synthesis in the future, namely, the union of agriculture and industry on the basis of the more perfected forms they have each acquired during their temporary separation.
>
> (Marx and Engels, 1996, 506)

No less than in a greening of industrial towns, the "higher synthesis" of the future union would undoubtedly result in the industrialization and urbanization of the country. Rather than Arcadia, we should probably imagine something closer to Owen's township communities, or Godin's Familistère, still extant and open for visit in the small town of Guise in northern France. William Morris himself called his "A Factory as It Might Be" (Morris, 1934, 647).

However appealing or distasteful we may find such a vision, we have to recognize that it is strikingly different from what most of us have been taught to expect from Marxism, even if no less profoundly transformative. For one thing, it makes Marx an early champion of Localism, one of the values increasingly espoused by many actors in the contemporary environmental movement. If this reading is justified, it has important implications for the question of limits we have so far found it difficult to decide in Marx's legacy, since a local economy must confront the existence of limits much more concretely than one reliant on global imports and distant outsourcing. It also makes Marx's ideal, he who was a lifelong city dweller and seemingly very much a town-animal, much more congenial to the wide current in environmentalism that shares his animus against enormous urban developments cut off from the natural world, even if the fusion it called for could hardly have been bucolic. On the other hand, it is a vision that is if anything even further from the reality of what happened historically than the received version of a productivist and Promethean socialism, and even less likely to be realized in the future.

To be sure, several spectacular achievements of late capitalism in technology and commodity availability have indeed erased the difference between town and country life in ways that conform to Marx's predictions, and perhaps hopes. It is not certain that the internet alone is sufficient to rescue anyone from idiocy, but in highly developed countries, it is making information just as accessible in the remotest rural areas as in urban centers, opening up possibilities we are only beginning to understand. More problematically, but no less surely, means of material exchange and individual transportation have made country life in the industrialized world just as full of creature comforts and close to most services as life in the cities, albeit at an energy and environmental cost that is proving unsustainable. And most dramatically, agriculture itself has been industrialized and transformed by science and technology to a degree that surpasses both Marx's highest expectations of a productive fusion with industry, and his worst fears of capitalism's destructiveness of the "lasting fertility of the soil." But far from mending the metabolic rift between town and country, and between human beings and nature, the accelerating process of modernization since Marx's time has on the contrary increased that rift and deepened the alienation of most city dwellers from the natural sources of their livelihood beyond anything he could have imagined. In the most advanced industrialized societies, the proportion of people actually engaged in agricultural production has now shrunk to very few percent of the local population[9] (while in many cases still able to produce a substantial surplus for exportation), and the production itself has become entirely dependent on the massive importation of artificial fertilizers and fossil fuels (not to mention migrant laborers).

It is true that recently we have begun witnessing, in reaction to this hyper-development in the industrialized world, a growing movement toward organic farming and more local production, still very small as a proportion of the overall output, but that may gain significant mainstream influence as the environmental damage of industrial agriculture is recognized as less and less sustainable. This might have pleased Marx, in a small way. No one in the real world, however, not even the greenest-eyed eco-dreamer, would now dare to imagine anything close to the fusion between town and country, and the abolition of our alienation from nature, that was his ultimate goal for the final stage. In the end, much as he tried to distance himself from the idealism and naivety of all social utopias, Marx's most explicit formulation of a dialectical resolution to the material conflict created by modernity, the closing of the metabolic rift itself, may also turn out to be his most utopian.

Compared with Marx's, John Stuart Mill's evocation of the final stage may first appear just as utopian, and it is if anything even less developed relative to the considerable body of writing also produced by Marx's contemporary. "The Stationary State," the only chapter explicitly devoted to this notion in Mill's oeuvre, runs a mere six pages in his two-volume *Principles* (first edition 1848). Furthermore, this shortest of all the chapters barely touches on the problem of economic and ecological limits in the form that we have encountered it so far, particularly in Marx. Mill does not even seem concerned to establish that we must, or whether we can, reconcile the needs of human progress with the limitedness of the natural world. Rather, and more radically, he displaces or bypasses the problem by promoting the acceptance of limits as a *preferable* condition for humanity. Mills is the first, and for a long time the only major thinker of modernity to envision the end of economic growth not as a limitation imposed on us by necessity, but as a good in itself, a goal to be embraced rather than feared, because it would be better for us, both socially and humanly.[10] In this he opens the way – quite consciously – to rethinking the question of limits out of the negativity in which modernity has continuously framed it, and in terms that, as we will more fully recognize later, are in many ways still unsurpassable for our own much delayed confrontation with the finitude of the world.

This does not mean that Mill, in his work as a whole and the *Principles* in particular, is unaware of the existence, and indeed of the problem of limits. The most evident manifestation of this consciousness is his ardent, and in some ways surprisingly unqualified support of Malthusianism. For Mill, rather kind-hearted himself and leaning increasingly toward socialism, goes out of his way to defend Malthus' law of wages against accusations of "hard-heartedness." He presents as "a passing remark of Mr. Malthus, hazarded chiefly by means of illustration" the famous arithmetical and geometrical ratios of food and population increase, deeming this "unlucky attempt … wholly superfluous to his argument" (Mill, 1848, 359). True enough in substance, but one would not think so from the prominence given the ratios by Malthus himself, at least in the first two editions of the *Essay*, and never wholly abandoned in succeeding ones. Similarly, Mills smoothes over as corrections of "mere language" rather important distinctions between an actual "tendency of population to *increase faster* than the means of subsistence" (emphasis his own), or whether "population *would* in most circumstances increase faster than the means of subsistence, *if it were not checked* either by mortality or by prudence" (emphasis added). In short Mill, who would have many good reasons to distance himself from Malthus, eschews every opportunity to do so, presenting himself as a loyal follower and defender.

But Malthusianism in Mill's hands, as for the advocates of the movement that would finally gain prominence under that name at the end of the nineteenth century, has really become something else altogether. Rather than a malediction visited upon the laboring masses, from whose natural law they could never be expected to escape (notwithstanding Malthus later allowing for the possibility of "preventive checks,") the principle of population is recognized by Mills as a legacy of cultural prejudice, perpetuated by the rich and among the poor by the combined forces of religion and "a spurious delicacy." As if children, he writes, "were rained down upon married people, direct from heaven, without their being art or part in the matter" (ibid., 374–375). The great numbers of waged laborers, he agrees, do contribute to their misery by bringing into the world more children than they can support, and more than will eventually find employment, thereby depressing wages even further, but this cycle can be broken, and reversed to their advantage. Thus what had been asserted by Malthus as the inevitability of an excessive population, forever pressing against the limits of the earth's natural carrying capacity, has been turned into the call for a massive campaign to change social attitudes toward procreation, and the militant advocacy of family planning and birth control for the sake of improving the condition of the poor.[11] Not necessarily the opposite of what Malthus intended, on a generous reading, but certainly not by the means he would have approved.

That the problem of population is for Mill essentially a social one, and therefore amenable to social solutions – a point which may be obvious to us, but was by no means universally accepted in the wake of the original Malthusian pronouncements – is made clear by his immediately linking it, in the chapter entitled "Remedies for low wages," with the question of public education, and with the crucial factor of the status of women. In both cases Mill's remarkably progressive agenda is explicitly based, as indeed all cultural progressivism must be, even when it is not so explicit, on a thoroughly un-Malthusian belief that moral dispositions and social attitudes, even those apparently natural or most deeply engrained in tradition, can be changed through the power of what he calls "opinion." "Those who think it hopeless," he writes, "that the labouring classes should be induced to practice a sufficient degree of prudence in regard to the increase of their families, because they have hitherto stopt short of that point, show an inability to estimate the ordinary principles of human action" (ibid., 376). They "can never have considered," he adds on the following page to insist on the most decisive of those principles, "how large a portion of the motives which induce the generality of men to take care even of their own interest, is derived *from regard for opinion*," a motivation he immediately and crucially explicates as "*– the expectation of being disliked or despised for not doing it*"[12] (ibid., 377).

One should probably refrain from over-theorizing what is for Mill simply a common-sense observation, but it is remarkable to see such an eminent representative of liberalism, supposedly steeped in utilitarianism, recognize that "even self-interest" is so often inseparable from our regard for the sentiments of others; that the self, for human beings, is almost always dependent on the other – that it is other-regarding. Or to use slightly more current language, that far from being purely monadic, egoistical agents, as a reductive version of economic psychology would have it, we are fundamentally intersubjective beings. And it is worth remembering that Mill is not the only one, nor the first among the great liberal economists to emphasize this dimension of human motivation. One need only rephrase Mill's gloss on the principle he calls "regard for opinion" in its complementary positive terms, as: "from the expectation of being liked and admired for doing it," to rediscover almost word for word the language used by Adam Smith to describe the operation of what he called "sympathy."[13] We are moved by our identification with the sentiments (or the opinion) of our peers, or those we admire most. For both advocates of liberalism, the self is primarily interested in the interest of the other.

It is also no coincidence that neither Mill nor Smith thought that liberal capitalism – the regime of economic liberty – could work, or could work in a way that was humanly acceptable, without the counterbalance of universal public education. For Smith, this was necessary to compensate for the stultifying effects of an ever-increasing division of labor (his version of alienation) on the vast majority of workers. For Mill, it is primarily to "cultivate common sense" (rather than reason, we might want to note) among the mass of the people; "to qualify them for forming a sound practical judgment of the circumstances by which they are surrounded" (Mill, 1848, 381). But education is also essential for Mill, in a way that it could not have been for pre-Malthusian Smith, to bring about a reversal in the commonly shared assumption about the desirability of having as many children as possible, based on the principled belief that "once opinion shall have turned itself into an adverse direction, a revolution will soon take place in this department of human conduct" (ibid., 378).

And to accomplish this "revolution" of opinion, Mill knows that he can especially count on the powerful support of "the great majority of women." Mill's justly recognized feminism is ever-present in his thought, and in the chapter we are following, it gives rise to one of his eloquent declarations of moral outrage at the subservient legal status of women: "Among the barbarisms which law and morals have not ceased to sanction, the most disgusting surely is, that any human being should be permitted to consider himself as having a *right* to the person of another." But this feminist consciousness is not just one of principle. For the question of population, it has eminently practical consequences. For Mill is well aware that it is women who bear "(along with all the physical suffering and at least a full share of privations) the whole of the intolerable domestic drudgery from the excess" of families that are too large. "To be relieved of it," he confidently foresees, "would be hailed as a blessing by multitudes of women who now never venture to urge such a claim, but who would urge it, if supported by the moral feelings of the community." There would therefore be no need, he concludes, to enforce anti-natalist policies though legal means "if women were admitted, as on all other grounds they have the clearest title to be, to the same rights of citizenship with men."

> Let them cease to be confined by custom to one physical function as their means of living and their source of influence, and they would have for the first time an equal voice with men in what concerns that function: and of all the improvements in reserve for mankind which it is now possible to foresee, none might be expected to be so fertile as this in almost every kind of moral and social benefit.
>
> (Ibid., 379)

One can only wonder at Mill's prescience (the last sentence was added in 1852), at the justness of his appreciation of what is surely turning out to be one the most transformative movements of late modernity, a feminist revolution "fertile" indeed "in almost every kind of moral benefit." His sense of history in that regard, purely based on moral conviction, and in this sense so un-Marxian, has passed the test of time incomparably better than that of the theorist of historical materialism. For we now know that there is no higher correlation in social behavior than that between the status of women, and in particular women's educational level, and the rate of fertility in a given population. In every country in the world where girls have finally gained access to schooling and the prospect of some self-determination, the number of children per family has fallen, or is in the process of falling to a simple reproduction level. World population will inevitably continue to grow for many decades, and it is already pressing dangerously against natural resources, but as the present social trends continue to extend to all cultures on every continent, we can hope that the numbers of the human race will reach self-limitation by the end of the present century. If Mill's vision can be said to be Malthusian, then we can say that Malthusianism is winning over the world.

But of course, not all limits are amenable to human will. Mill does in fact draw a clear distinction between productions that are subject to an eventual (even if only theoretical) limitation, and those whose potential increase is inherently infinite. Among the former he naturally places agriculture, but also the mining of "almost all kinds of raw material extracted from the interior of the earth – metal, coals, precious stones, &c." (ibid., 474). These are classically ruled by the law of diminishing returns: "the law," as he puts it, "that increased labour, in any given state of agricultural skill, is attended with a less than proportional increase of produce." They generate differential rents as it becomes necessary to exploit lands of lesser fertility, and they are eventually (again, theoretically) "either wholly or partially exhaustible; such as coal, and most if not all metals; for even iron," he adds, "the most abundant as well as the most useful of metallic products, which forms an ingredient of most minerals and of almost all rocks, is susceptible of exhaustion so far as regards its richest and most tractable ores" (ibid., 702).

On the other hand, for Mill industrial production, what he still calls "manufacturing," is not in and of itself subject to such limitations. It is on the contrary the beneficiary of the opposite tendency toward what we have come to call economies of scale: "The larger the scale on which manufacturing operations are carried on, the more cheaply they can in general be performed." From this he can neatly conclude:

> The tendency, then, being to a perpetual increase of the productive power of labour in manufactures, while in agriculture and mining there is a conflict between two tendencies, the one towards an increase in productive power, the other toward a diminution of it.
>
> (Ibid., 703)

A little too neatly perhaps, since as he had just granted in the preceding paragraph: "As manufacturers, however, depend for their materials either upon agriculture, or mining, or the spontaneous products of the earth [all of which, we recall "depending on a material which is not renewed, are either wholly or partially exhaustible,"] manufacturing industry is subject, in respect to one of its essentials, to the same law as agriculture." That is to say that manufacturing too, to the extent that it is always dependent on non-renewable materials (and, we would add, on non-renewable energy), must eventually be subject to the same law of limits that governs all material production (except one that would be entirely renewable) in a finite world.

We can therefore find in Mill the analytical framework for the theoretical possibility of human production reaching natural limits, a possibility that would make it necessary to constrain either population or the consumption of goods, and force us to renounce economic growth. Except that it is a possibility Mill does not actually believe in, and that he even goes so far as to deny in practice. It is "a contingency," he writes in the concluding sentence of the chapter on rents, "which the almost inevitable progress of human culture and improvement in the long interval which has first to elapse, forbids us to consider as probable" (ibid., 477). Quite to the contrary, we find him affirming in a familiar argument in the chapter on "the progressive state of wealth":

> The progress which is to be expected in the physical sciences and arts, combined with the greater security of property, and greater freedom in disposing of it [...] afford space and scope for an indefinite increase of capital and production, and for the increase of population which is its ordinary accompaniment. That the growth of population will overpass the increase of production, there is not much reason to apprehend; and that it should even keep pace with it, is inconsistent with the supposition of any real improvement in the poorest classes of the people.
>
> (Ibid., 699)

Even about the potential limits of agricultural production, the traditional rationale for Malthusianism, he refuses the dire predictions of the pessimists. "We can imagine," he concedes, "a country so fully peopled, and with its cultivable soil so completely occupied, that to produce any additional quantity would require more labour than the produce would feed." We can even "suppose this to be the condition of the whole world." If it were so, and population kept increasing, then indeed "both the land and its produce would really rise to a monopoly or scarcity price." But just as Adam Smith had done when contemplating such a theoretical hypothesis, Mill ends up rejecting it as a likely possibility: "But this state of things never can have really existed anywhere, unless possibly in some small island cut off from the rest of the world, nor is there any danger whatever that it should exist." So long as there is any surplus of cultivable land, as is the case, he affirms, "in all countries of any extent," then "it is the same thing, so far as that quantity of land is concerned, as if there were an infinite quantity" (ibid., 472–473).

It is therefore all the more striking to find Mill reintroducing, in the chapter "Of the Stationary State," the idea of limits as the natural goal of economic progress, as an "ultimate point" toward which the progressive movement of society, "not in its nature unlimited" is inevitably tending. For "it must always have been seen [...] by political economists," he now claims, "that the increase of wealth is not boundless: that at the end of what they term the progressive state lies the stationary state, that all progress in wealth is but a postponement of this, and that each step in advance is an approach to it" (ibid., 746). But the concept of limits that makes it possible to think simultaneously the imminence of the goal and its infinite postponement owes something to the famous pre-Socratic paradox of the arrow never reaching the target. "This ultimate goal," in Mill's modern version, "is at all times near enough to be fully into view," "we are always on the verge of it," but – and this is the crucial point: "if we have not reached it long ago, it is because the goal itself flies before us." The arrow fails to reach its goal only because the latter is a moving target; and it would undoubtedly reach this goal if the target stopped moving. But then the paradox threatens to become a mere tautology. Of course, we want to comment without irony, a (progressive) economy would soon become stationary if it stopped progressing. "The richest and more prosperous countries," as Mill himself puts it just as seriously, "would very soon attain the stationary state, if no further improvements were made in the productive arts." But it is in the very nature of a progressive economy to constantly push away its limits through "further improvements … in the productive arts," and to forever expect, as we have seen Mill affirm just a moment ago, that "progress … in the physical sciences and arts" would postpone the attainment of those limits. From Adam Smith's indefinite extension of the division of labor to Joseph Schumpeter's creative destruction, the quantitative expansion of productive capacity has always been understood to rely on a qualitative improvement of the means of production. It is only because of innovation and technological invention that the economy remains progressive, that it manages to avoid reaching the stationary state toward which it naturally tends. The progressive state is indeed a *fuite en avant*, a race against its own impending finitude.

Thus, in spite of his pronouncements on limits at the beginning of the chapter on the stationary state, Mill's true originality is not as a prophet of doom, warning of the dangers of overreaching the carrying capacity of the natural world – which would make him, say, an early adherent to the Club of Rome. He is in fact much less prescient than Marx in this regard, since he seems to think that, for a long time if not indefinitely, human ingeniousness and creativity *could*, if we chose to, delay what is at first presented as an inevitable day of reckoning. His more radical argument is not that we cannot forever win the race (we just might be able to), but that we would all be better off if we stopped racing – if we really accepted the stationary state as the goal, as the ultimate achievement of human progress, rather than its failure. In this, he is astonishingly original among classical economists, and quite conscious of being so. The "impossibility of ultimately avoiding the stationary state," he remarks in what is if anything an understatement, "must have been, to the political economists of the last two generations, an unpleasing and discouraging prospect; for the tone and tendency of their speculations goes completely to identify all that is economically desirable with the progressive state." Adam Smith himself, he justly reminds us, "always assumes that the condition of the mass of the people, though it may not be positively distressed, must be pinched and stinted in a stationary condition of wealth [Smith's actual words were "hard" and "dull"], and can only be satisfactory in a progressive state" (ibid., 747). And such will remain, we could add, the assumption shared by all the following generations of political economists to our own day. Underlying the debate between the pessimists and the optimists, that is to say those who foresee that some future limit to growth is inevitable, and those who trust that it can always be postponed, the universal point of agreement is that the eventuality of reaching a stationary state is a dismal prospect to be avoided at all costs. A good economy, indeed, can only be in a progressive state.

It is this very assumption Mill sets out to challenge. "I cannot," he declares at the end of his assessment of the field, "regard the stationary state of capital and wealth with the unaffected aversion so generally manifested towards it by political economists of the old school." Quite to the contrary, he then goes on: "I am inclined to believe that it would be, on the whole, *a very considerable improvement on our present condition*"[14] (ibid., 748). We are so used to the stridency, and occasionally the brilliance, of revolutionary discourse that Mill's calmly measured, ever so politely bourgeois tone can easily make us miss the radicalism of his proposition. Not only that the stationary stage (the end of economic growth), can be contemplated without fear; that it need not be thought of as a depressing state, but that it would be *much better* than "our present condition" (mid nineteenth-century Britain) – that is to say the beginning of one of the most impressive periods of economic growth the world had ever known until the *Trente Glorieuses* and the rise of the Asian Tigers. And on what ground is Mill moved to reverse such an obvious advantage of the progressive over the stationary one? Not out of any urgent sense of limits, nor primarily out of concern for the preservation of the environment (although his appreciation for the value of the natural world will clearly appear at the end of the essay), but in the name of a different idea of what constitutes the good life, of an ideal of social fulfillment no less humanistic than that espoused by Marx in his more utopian moments, and ultimately, no more economically nor scientifically based than this ideal was for Marx himself:

> I confess I am not charmed with the ideal of life held out by those who think that the normal state of human beings is that of struggling to get on; that the trampling, crushing, elbowing, and treading on each other's heels, which form the existing type of social life, are the most desirable lot of human kind, or anything but the disagreeable symptoms of one of the phases of industrial progress.
>
> (Ibid., 748)

Mill's manifesto is at its core a repudiation of the rat race, of the endless pursuit of ever more riches to establish one's superiority over others. The "best state for human nature," he affirms instead: "is that in which, while no one is poor, no one desires to be richer, nor has any reason to fear being thrust back by the efforts of others to push themselves forward" (ibid., 748–749).

Thus for the first time in modernity – and for a long time in salient solitude – do we see a major liberal, progressive economist come without equivocation to the only conclusion a genuine acceptance of limits must always lead to: that the inevitable questioning of the desire for material accumulation as a social good is inextricably linked to the moral necessity of material equality among human beings. The one cannot be thought without the other (though of course both can be dismissed as impossibilities, jointly or separately, but more frequently the one in the name of the other). In Mill's advocacy for the stationary state, the virtue of restraint (or the rejection of the pursuit of wealth for its own sake) is never to be dissociated from the good of equality, and when one is temporarily emphasized, it is always with the goal of achieving the other, as its only condition of acceptability in a world of limits. If those, says Mill (and those of us, we want to add)

> who do not accept the present [...] stage of human improvement as its ultimate type, may be excused for being comparatively indifferent to the kind of economic progress which excites the congratulations of ordinary politicians; the mere increase of production and accumulation.

Or if we, as he again insists, "know not why it should be a matter of congratulation that persons who are already richer than any one needs to be, should have doubled their means of consuming things which give little or no pleasure except as representative of wealth," it is out of the conviction that in the "most advanced" countries "what is economically needed is a better distribution" of wealth (ibid., 749), not the endless pursuit of more riches. The ultimate goal of progress, the good to be posited as the end of growth, can only be based on balance and equity, contentment and equality. The stationary state must be conceived, both socially and economically, as a condition of static equilibrium.

One obvious question, and for many of us, a serious objection to Mill's conception of the last stage, is whether such a sweeping rejection of the "struggle to get on," of the competitive motive itself, is still compatible with capitalism. Isn't competition, so scornfully described by Mill (in a language that could easily have been used by Adam Smith) as the "trampling, elbowing, and treading on each other's heels," the principal motivation, indeed the very active principle of a liberal economy? With this sweeping critique of the individual "desire to better our condition" that had always been seen as the universal motor of economic progress, Mill seems to have reached the limits of the liberal model he had sought all his life to defend, promote and humanize, perhaps to the point of contradiction. It is no wonder that in the end he was more and more drawn to forms of associative socialism that, while preserving economic freedom, favored cooperation over competition. Whether this later ideal of a cooperative economy, which had widespread appeal in Europe for a long period of time spanning the second half of the nineteenth century and the first half of the following, particularly in Marx's England (although this was later obscured by the much greater publicity and aggressiveness of the Marxist movement) – whether such an alternative could have had a credible chance of surviving the double onslaught of capitalism and communism, we will never know. But the larger question of the compatibility of a viable stationary state with the exigencies of a capitalist economy, or whether capitalism can do without growth – a question paused by Mill perhaps unwittingly – remains an open and crucial one for anyone thinking within the horizon of finitude, and will necessarily return with our own confrontation with limits.

And yet, in spite of this irresolution, Mills is surely right, at least poetically if not economically, to insist that in all metaphorical logic the progressive state that we now call *growth* can only be "one of the phases" of economic progress, not the "normal state of human beings"; and that the "struggling to get on" is but one of its "disagreeable symptoms." It may be, he recognizes, "a necessary stage in the progress of civilization," and one that all cultures will eventually have to go through to reach maturity, but it should only properly be seen, as he perfectly puts it, as "*an incident of growth*"[15] (ibid., 748). The idea of economic growth in the sense of an increase in the wealth of nations, and of course the desire for the riches, of the household or of the city, to "grow," are at least as old as *oeconomia* itself. And on the other hand the intuitive comparison of the body politic, of the population under a sovereign head, of its strength and size, youth and maturity, with that of a living body, was already common in the early-modern reflections on the state from the sixteenth century.[16] As for the physiological analogy of a body *economic*, with its proper functions and condition for health, it was explicit as early as the writings of Quesnay, the father of the Physiocrats, "who was himself a physician," as Adam Smith was careful to note when referring to him in *The Wealth of Nations* (Smith, 1776, II, 674). But the use of the term *growth* itself to speak of an expanding economy, so naturalized today that it is understood as such by the least economically inclined, came surprisingly late in the history of the discipline, and it was still not common in the nineteenth century. And so when Mill employs the word to describe the progressive state as "an incident of growth," it still has all the power of an uneroded metaphor, because it *is* a metaphor for Mill, to be taken seriously in all its implications. The most evident of these, if largely ignored in modern usage, is that a growing economy is still immature, it is not yet fully grown up; even if no longer infantile, it is still young. Growth cannot therefore be the normal, healthy condition of a fully developed, adult economy, nor the ultimate aim of social life. It may indeed be "a necessary stage in the progress of civilization," but it cannot be its ultimate goal for maturity. Growth, the state we have come to expect in the last 200 years as the normal condition of a healthy economy, should instead be seen, Mill seems to suggest in a long-term perspective, as an incident of civilization.

Another, perhaps even more difficult question raised by Mill's short manifesto in favor of the stationary state – it still is the most formidable objection to any argument for the end of growth – is that of the persistence of poverty in the world, even if Mill's position on this subject too is in the end theoretically unimpeachable. On the one hand it is a remarkable lesson in relativism to find that in the middle of the nineteenth century, over 150 years ago, a period we now see as only the beginning of large-scale industrialization and rapid growth, Mill considers that "in the most advanced" countries, the middle and upper classes are already "richer than any one needs to be," and does not see "as a matter of congratulation" that they "should have doubled their means of consuming things." Since that time, the level of material consumption among the corresponding social classes in advanced countries has of course much more than doubled again, and many still do not think it is high enough. Surely not a matter of approval for Mill, who as we saw advocated instead for "a better distribution." On *that* score however – the degree to which resources are spread out within an economy – even if the record is mixed and current trends regressive, some appreciation could be expected from him if he were able to witness the evolution of the gap between rich and poor during the same period of time. For in those same advanced countries, especially those that have developed strong systems of public services and social redistribution, the standard of living of ordinary people (if not their nominal wealth) have progressed considerably in relation to that of the most privileged, however extravagant the latter may still sometimes be. Indeed, to an extent that would have astounded a mid-nineteenth century observer. To be sure, as the long term analyses of Thomas Piketty and others have shown, inequalities have again been rising alarmingly since the 1980s, but up until the previous decades developed countries had achieved a narrowing between the real living conditions of the rich and the poor that was probably unprecedented since humanity left the hunting and gathering stage.[17] Even those of us who feel that this progress toward social and economic equality has not gone nearly far enough, and worry that it is being threatened, should acknowledge that it is one of the most meaningful achievements of modernity (albeit one that is no longer environmentally sustainable in the manner in which it has been achieved).

But the question of distribution, of alleviating the gap between rich and poor, is not only a central concern within each national economy, but much more profoundly, and in historical terms, tragically, between countries in different parts of the world. Mill does recognize this historical inequality, and therefore the need for poorer areas of the world to keep growing economically while the most affluent remain stationary, until such a time, we can implicitly and hopefully infer, when all shall be on the same level. "It is only in the backward countries of the world," as he puts it, "that increased production is still an important object." On a generous reading, this is the right position to take, and the latter part of the proposition at least can hardly be objected to by anyone concerned with economic justice. But even the most kindly disposed contemporary reader must find this sole comment on such a fundamental problem wanting. It isn't just the formulation itself that can be judged offensive. In this respect Mill is certainly no worse than the majority of *his* contemporaries. In spite of the inevitable resonance of "backward" to our ears, Mill always resisted the racism so many accepted at the time as scientific doxa. He is simply voicing his unquestioning post-Enlightenment belief in the superior achievements and social conditions of Western civilization and those parts of the world that had already been brought under its influence, and the assumption that of course the rest must eventually benefit from the same material, scientific, and political progress. We would undoubtedly prefer a more qualified, and culturally sensitive position on the part of Mill, but he does not deserve to be demonized for it.

But neither can he be for us a model, let alone a hero on this score. It is well known that while employed for many years by the East India Company – and so directly involved in and benefiting from Britain's colonial business with India – Mill never bothered to go there, or anywhere else outside of Europe for that matter. Had he been directly exposed to the reality of what we now call developing nations (among which India, the country he must at least have been informed about, never was by any means the poorest), it is doubtful that he would have so summarily disposed of the immense gap to be bridged between the advanced and the "backward" countries. It is true that at the time he wrote, with world population as it was in the poorer parts of the world, and at the levels of material consumption then prevalent in the richest, it might still have been possible for the poorest to catch up with the rich within the resources available in the world and without overwhelming the carrying capacity of the earth, a scenario that has since been rendered utterly impossible by the continued growth of both population and material consumption. And to his credit, Mill had the rare foresight, or moral compass, to oppose these trends, especially the high birth rates among the poor, in which he saw the source of their continued immiseration, but also, as we just saw, the continued enrichment of the already needlessly rich. The general direction suggested by his positions was right, but one wishes he had done much more than merely point to it, almost in passing, or at least acknowledged how much the world would have to change to approach a semblance of equality, and how difficult such a change would be to achieve. In this respect also, Mill is perhaps most relevant by the questions he raises seemingly unawares, and therefore leaves without answers.

This is not the last word of the essay, however, and in the development that actually ends it, the last two long paragraphs, Mills goes a long way toward redeeming his earlier failings by revealing an astonishingly modern sensibility, and a wisdom seldom remembered in modernity. The opening subject is once again population, together with inequality the most enduring of Mill's preoccupations. And there again he poses the question of limits, only to lead to a dimension of human experience, and to an appreciation of the natural world that we had not encountered before, in him or any other major thinker on the subject. "There is room in the world, no doubt," Mill concedes, "even in old countries, for a great increase in population, supposing the arts of life to go on improving, and capital to increase" (Mill, 1848, 750). As we have seen before, the questioning of growth, for Mill, is not primarily driven by a sense of obligation or urgency, by the conviction many of us have now that we have already reached absolute limits. Population, in his estimation, could keep getting larger, even in the "most populous" countries, perhaps for long time. But is it a good thing that it should do so? "I see very little reason for desiring it," Mill answers. Again, we find him taking a principled position against the maximum increase possible, as we had seen in his observations on wealth and consumption, not however for the sake of equality or as a rejection of endless competition, but this time in the name of another "good" to be valued beside these: solitude. For "it is not good for man," he reflects, "to be kept perforce at all times in the presence of his species" (ibid., 750). Solitude, therefore, is good for human beings, and conversely, "a world from which solitude has been extirpated is a very poor ideal." But there are actually two senses in which solitude is good for us in Mill's proposition, both of which carry considerable cultural value.

Solitude, first, as the opportunity for all human beings to be sometimes by themselves. "Solitude, in the sense of being often alone," Mill affirms, "is essential to any depth of meditation or of character." A sentiment many of us in the modern, developed, and crowded world would undoubtedly greet with a great deal of approbation and sympathy. But would it be understood, let alone approved of, by most if not all human beings sharing the planet today (not to mention our ancestors from a more or less distant past)? Different cultures in all parts of the world have traditionally prescribed and valued certain experiences of solitude for certain of their members at certain times of their lives, for enlightenment, or penance, as a test of fortitude, or a condition for reaching enhanced mental states, though perhaps not more often than as a form of punishment, as in solitary confinement or banishment. But the idea that everyone, in the ordinary course of life, needs to be "often alone" for his or her wellbeing, that it is indeed a right as well as a privilege to be valued above the mere peopling of the earth, that view would surely not have been recognizable before the development in the West of a modernity founded on the respect of individual personhood, before what we might call the democratization of privacy. There are still many people in the world today who almost never have such an experience, and do not ordinarily feel a need for it, if they do not actually dread it. Even among the many hyper-modern inhabitants of the most densely populated megapolises of the world, there are very great cultural and individual variations in the value accorded to personal space, and the desirability of being by oneself. While this value of solitude, so strikingly singled out by Mill as a good so essential that it should take precedence over absolute limits, will instantly resonate with many of us, perhaps most often in the West, it is not certain that it would be given nearly the same priority, let alone be treasured as a universal need, by most people in the world today.

But the other value of solitude emphasized by Mill is perhaps even more culturally bound. If we can draw an analogy with a more famous distinction, this second sense of solitude is not merely solitude *from* other human beings, but solitude *to* experience nature itself as a good.[18] For Mill "solitude in the presence of natural beauty and grandeur, is the cradle of thoughts and aspirations which are not only good for the individual, but which society could ill do without." Hence the very modern need (perhaps already an unrealistic ideal) to preserve a part of the natural world from the encroachments of civilization. There would not be, Mill foreshadows:

> much satisfaction in contemplating the world with nothing left to the spontaneous activity of nature; with every rood of land brought into cultivation, which is capable of growing food for human beings; every flowery waste or natural pasture ploughed up, all quadrupeds or birds which are not domesticated for man's use exterminated as his rivals for food, every hedgerow or superfluous tree rooted out, and scarcely a place left where a wild shrub or

> flower could grow without being eradicated as a weed in the name of improved agriculture.
>
> (Ibid., 750)

The sentiment expressed so strikingly and so early by Mill has since then been embraced by a very great many people in the world, and countless local, national and international institutions. It is explicitly at the root of the wilderness conservation movement that started in the early twentieth century in North America, to spread, unevenly, to many parts of the world, and without being always so clearly articulated, it is undoubtedly shared by very large segments of those who identify with environmentalism and the protection of nature. It can even be said without exaggeration that for growing numbers of our contemporaries, as it was long ago for all of our ancestors, communion with the natural world is once again becoming a form of quasi-religious experience, the last refuge of the sacred.

And yet such a reverential attitude is far from universal. In the sensibility of the West itself, it would have been all but incomprehensible before the Romantic turn of the late eighteenth century Mill is clearly heir to, notwithstanding his association with utilitarianism. In his own days, a pragmatic, purely instrumental relation to nature was still largely prevalent in deed if not in thought, and we have seen that Marx's vision of an eventual merging of town and country seemed wholly indifferent, if not antagonistic, to the value of preserving pristine natural beauty for human contemplation and moral inspiration. In the contemporary global debate on environmental resources, the conflict between the importance of conserving the last wilderness areas and primeval forests on earth and the pressing survival needs of the populations living in or near them has become one of the most painfully contentious issues of political ecology. The desperate call many of us feel to save an invaluable and irreplaceable natural legacy from unconscionable destruction – from what is surely nothing less than a crime against the universe – is to others nothing more than the romantic luxury of the privileged. The "sincere hope" with which Mill concluded his defense of solitude in the world, the hope that "long before necessity compels them to do so," future generations "will be content to be stationary" rather than deprive the earth of "that great portion of its pleasantness which it owes to things that the unlimited increase of wealth and population would extirpate from it" – that hope can surely not be assumed to be shared by everyone, or ever to be reached, if it is not already too late.

What Mill's most modern, and to some of our contemporaries, most affecting plea for the restraint of growth reveals perhaps most clearly is that the question of limits, already so difficult to confront in its seemingly undeniable physical forms, is also rarely separable from the values different individuals and human communities hold to be dearest. If some limits can be shown to be almost as solidly objective as a wall – the range of temperatures hospitable to human life on earth; the critical threshold beneath which a species will become instinct; even, with a greater degree of uncertainty, the reserves of usable fossil fuels –

many others are in the eye, or the heart, or the morality of the beholder. Science may succeed in persuading the great majority of human beings that a certain proportion of microorganisms in the oceans, even if invisible, is crucial to the health of the biosphere, and therefore a limit not to be trespassed, but it will probably never convince everyone that the survival of a unique frog, or a tiny inedible fish in a featureless marshland, is more important than extending a rice-field, or even developing a marina. Behind the question of hard limits, already hard enough to think through, though we should try to as objectively as possible, what should eventually be confronted, as Mill was perhaps the first to recognize, is the question of goods. But first, we have yet to bring to the present the question of limits itself.

Notes

1 This is perhaps not quite true. Alfred Marshall (1842–1924) was a major economist, and although looking forward to the neo-classical school of the following century, he did live the great majority of his life in the nineteenth, and published his *Principles of Economics* in 1890. The question is whether he can be said to have embraced a version of the stationary state. W. W. Rostow cites this appealing passage from the *Principles*:

> The world would be much better if everyone would buy fewer and simpler things, and would take trouble in selecting them for their real beauty; being careful of course to get good value in return for his outlay, but preferring to buy a few things made well by highly paid labour rather than many made badly by low paid labour.
> (Marshall, 1930, 137)

And Rostow comments: "This is not quite Mill's stationary state with high quality of life, but bears a family relationship to it." When assessing Marshall's position with regards to limits, however, he is more conclusive:

> As for the long-term prospect for growth in the world economy as a whole, Marshall was [...] something of a limits-to-growth pessimist. Looking back and projecting forward [...] his sense was that, despite the potentialities of science, diminishing returns to natural resources would constrain the expansion of the world economy [...], and he concluded that this constraint would prevail before the end of the twenty-first century.
> (Rostow, 1990, 167, 188–189)

2 Auguste Comte's *Cours de philosophie positive* (1830–1842), Johann Jakob Bachofen's *Mutterrecht (Mother Right*, 1861) and Lewis Morgan's *Ancient Society* (1877) are all examples of diversely influential works based on a theory of progress through stages of historical development.

3 The *Phalanstère* is Charles Fourier's detailed design for a utopian commune, as theorized in *La théorie des quatre mouvements*, 1808 (Fourier, 1996). For a general history of utopias, see Frank and Fritzie Manuel: *Utopian Thought in the Western World* (Manuel and Manuel, 1979).

4 See notably the two studies that have been the principal guides for my own discovery of the environmental Marx, Paul Burkett's *Marxism and Ecological Economics* (Burkett, 2006) and John Bellamy Foster's *Marx's Ecology* (Foster, 2000).

5 Remembering that Jevons' *The Coal Question* was first published in 1865.

6 *Klima und Planzenwelt in der Zeit, eine Geschischte beider* (Landshut, 1847).

7 Although they do not claim a Marxist heritage, this is also clearly the dominant thesis of Kent Flannery and Joyce Marcus in their powerful *The Creation of Inequality*:

How Our Ancestors Set the Stage for Monarchy, Slavery, and Empire (Flannery and Marcus, 2012).

8 William Bellamy Foster, who wants to argue strongly for an ecological Marxism, nevertheless acknowledges the dominant criticism of the "productivist" and "Promethean" premise of Marxism, represented by Ted Benton, Reiner Grundman, and Anthony Giddens, among others (Foster, 2000, 134–135). Victor Ferkiss summarizes this view: "Marx's attitude toward the world always retained that Promethean thrust, glorifying the human conquest of nature" (Ferkiss, 1993, 108). John Clark, also quoted by Foster, is even more absolute in his judgment:

> Marx's Promethean ... "Man" is a being who is not at home in nature, who does not see the earth as the "household" of ecology. He is an indominable spirit who must subject nature in his quest for self-realization ... For such a being, the forces of nature [...] must be subdued.
>
> (Clark, 1989, 258)

Paul Burkett, adding James O'Connor and Juan Martinez-Alier to the roster of ecological critics of Marx, reaches the same conclusion:

> Indeed, the conventional wisdom among eco-socialists now seems to be that classical Marxism, as represented by Marx and Engels, is seriously flawed ecologically unless augmented with, and even largely supplanted by, a "Polanyian" analysis of the conditions of production together with a certain element of green theory.
>
> (Burkett, 2006, 7)

9 According to data collated by the International Labour Organization from 2017, and reported by the World Bank, the percentage of people employed in agriculture was 3 percent in High Income countries and in the Euro area, 2 percent in the US, Canada and the Netherlands, and 1 percent in the UK and Germany (World Bank, 2017).

10 The next comparable statement of preference for economic appeasement, barely a few pages longer, is John Maynard Keynes' "Economic Possibilities for Our Grandchildren" (1930), which is also garnering much late attention (Keynes, 2009, 192–202).

11 The young John Stuart Mill was once arrested for distributing birth control literature. Given his birth date of 1806, this would have been the time when Francis Place had begun distributing pamphlets to the poor, around 1823 (Mayhew, 2014, 134; Himes, 1963, 213).

12 Emphasis added.

13 *The Theory of Moral Sentiments*, especially Chapter II of Book III: "Of the love of Praise, and of that of Praise-worthiness; and the dread of Blame, and of that of Blame-Worthiness," which begins: "Man naturally desires, not only to be loved, but to be lovely; or to be that thing which is the natural and proper object of love" (Smith, 1759, 113). In the section "Of the Character of Virtue" in Book VI, Smith writes: "The desire of becoming the proper objects of [the respect of our equals...] is perhaps the strongest of all our desires..." (ibid., 213). And perhaps most radically, in the section "Of the Selfish Passions" in Book I: "... the chief part of human happiness arises from the consciousness of being beloved" (ibid., 41).

14 Emphasis added.

15 Emphasis added.

16 Witness Chapter 24 of Hobbes' *Leviathan*: "Of the Nutrition, and Procreation of a Commonwealth" (Hobbes, 1651).

17 It is also true that, according to the same analyses, this was not achieved only through intentional social policies, but also because of the devastating consequences on the *rentier* class of the two World Wars and the Great Depression (Piketty, 2014). This is not encouraging for long term systemic trends in the absence of political redress, as also shown by Emmanuel Saez, for example in "Striking it Richer" (Saez, 2015).

18 The distinction is drawn from Isaiah Berlin's famous "Two Concepts of Liberty." Its application to the concept of solitude, however, is not to be found in Berlin's essay (Berlin, 2002).

References

Bachofen, Johann Jacob (1992) *Myth, Religion, and Mother Right*. Princeton, NJ: Princeton University Press.

Benton, Ted (1995) "Adam Smith and the Limits to Growth," in *Adam Smith's Wealth of Nations*, Stephen Copley and Kathryn Sutherland (eds.). Manchester and New York: Manchester University Press.

Berlin, Isaiah (2002) *Liberty: Incorporating Four Essays on Liberty*. Oxford: Oxford University Press.

Burkett, Paul (2006) *Marxism and Ecological Economics: Toward a Red and Green Political Economy*. Leiden: Brill.

Clark, John (1989) "Marx's Inorganic Body," *Environmental Ethics*, vol. 11, no. 3, 243–258.

Diamond, Jared (2011) *Collapse: How Societies Choose to Fail or Succeed*. New York: Penguin.

Elster, Jon (1985) *Making Sense of Marx*. Cambridge: Cambridge University Press.

Engels, Frederick (1872) *The Housing Question*. Moscow: Progress Publishers (this edition 1975).

Ferkiss, Victor (1993) *Nature, Technology and Society*. New York: New York University Press.

Flannery, Kent and Joyce Marcus (2012) *The Creation of Inequality: How Our Ancestors Set the Stage for Monarchy, Slavery, and Empire*. Cambridge, MA: Harvard University Press.

Foster, John Bellamy (2000) *Marx's Ecology: Materialism and Nature*. New York: Monthly Review Press.

Fourier, Charles (1996) *The Theory of the Four Movements*. Cambridge, UK: Cambridge University Press.

Hegel, G. W. F. (1837) *Lectures on the Philosophy of History*. Aalten, The Netherlands: Wordbridge Publishing (this edition 2011).

Himes, Norman E. (1963) *Medical History of Contraception*. New York: Gamut Press, Inc.

Hobbes, Thomas (1651) *Leviathan, or the Matter, Forme and Power of a Commonwealth Ecclesiastical and Civil*. New York: Collier Books (this edition 1962).

Jevons, William Stanley (1905) *The Coal Question*. New York: Augustus Kelley.

Keynes, John Maynard (2009) *Essays in Persuasion*. New York: Classic House Books.

Lenzer, Gertrud (ed.) (2017) *Auguste Comte and Positivism: The Essential Writings*. London and New York: Routledge.

Manuel, Frank E. and Fritzie P. Manuel (1979) *Utopian Thought in the Western World*. Cambridge, MA: Harvard University Press.

Marshall, Alfred (1930) *Principles of Economics*, 8th edition. London: Macmillan.

Marx, Karl (1867) *Capital*. New York: Vintage (this edition 1981).

Marx, Karl (1974) *Early Writings*. New York: Vintage.

Marx, Karl and Frederick Engels (1878) *Anti-Dühring*. Moscow: Progress Publishers (this edition 1969).

Marx, Karl and Frederick Engels (1996) *Collected Works*. New York: International Publishers.

Mayhew, Robert J. (2014) *Malthus: The Life and Legacy of an Untimely Prophet*. Cambridge, MA and London: Cambridge University Press.

Mill, John Stuart (1848) *Principles of Political Economy, with Some of Their Applications to Social Philosophy*. New York: Augustus M. Kelley (this edition 1965).

Morgan, Lewis Henri (1985) *Ancient Society*. Tucson, AZ: Arizona University Press.

Morris, William (1934) *Selected Writings*. New York: Random House.

Piketty, Thomas (2014) *Capital in the Twenty-First Century*. Cambridge, MA: Harvard University Press.

Polanyi, Karl (1968) *Primitive, Archaic and Modern Economies: Essays of Karl Polanyi*. Garden City: Anchor.

Rostow, W. W. (1990) *Theorists of Economic Growth from David Hume to the Present: With a Perspective on the Next Century*. New York and Oxford: Oxford University Press.

Saez, Emmanuel (2015) "The Evolution of Top Incomes in the United States" [online]. Available at https://eml.berkeley.edu/~saez/saez-UStopincomes-2015.pdf.

Smith, Adam (1759) *The Theory of Moral Sentiments*. Oxford: Oxford University Press (this edition 1976).

Smith, Adam (1776) *An Inquiry into the Nature and Causes of the Wealth of Nations*. Oxford: Oxford University Press (this edition 1976).

Tainter, Joseph (1988) *The Collapse of Complex Societies*. Cambridge, UK: Cambridge University Press.

Tucker, Robert C. (ed.) (1978) *The Marx-Engels Reader*. New York, London: W. W. Norton.

World Bank (2017) "Employment in Agriculture Statistics" [online]. Available at https://data.worldbank.org/indicator/SL.AGR.EMPL.ZS.

5 The return of limits

In retrospect, and even in his own time, it is all too obvious that John Stuart Mill's call for the stationary state had as little chance of being heard, or at any rate of having a significant effect, as Karl Marx's vision of a reconciliation between town and country, except among a small minority of radical thinkers, or dreamers. A measure of how profoundly untimely his call really was is that it did not come anywhere near the end, but near the *beginning* of what we now know to have been the greatest economic expansion in the history of humanity, just as Marx was witnessing barely the onset of the great worldwide exodus from rural settlements to eventually dominant cities everywhere. The great transformation of modernity, we cannot help thinking, had to take its course, no matter how eloquent and passionate the warnings against its negative consequences, or the appeal of alternative conceptions of progress.

Another, sad if unsurprising lesson, is that the argument from wisdom, as we might call it, in counter-distinction to the argument from necessity, will probably never hold enough sway in the absence of necessity itself. We will not get wise unless we have to. Not until we have exhausted a vital resource, recent history seems to tell us, or until we are threatening the very conditions of our survival – when we are not just pressing upon limits, but in what is now often called "overshoot" – do we begin to envisage how we might do without that particular resource, or refrain from our threatening behavior. Necessity will perhaps always precede wisdom. But on the other hand, when we are truly persuaded that reserve or restraint is necessary (as Mill himself was not entirely), then the arguments from wisdom, those offered by Mill in the nineteenth century, but also those long cultivated by the many traditions of what constitutes the good life, contentment, and happiness for human beings, all the reflections on the immateriality of desire and fulfillment, may once again be valued for helping us see that the changes and renunciations imposed by necessity can also turn out to be beneficial for us, and in fact make us happier. We may be led to wisdom by necessity, but wisdom allows us to see necessity as a good. Indeed, wisdom itself may be nothing else than the recognition of the goodness of necessity.

This being said, who would now wish that Mill's call had really been heeded in his time? How can we possibly regret the extraordinary advance in the well-being and material standards of so many people in the world that continued economic growth has made possible since then, let alone the miraculous progress in science and knowledge that has been both its condition and its reward? Can we even imagine (supposing there is any sense in engaging in such retrospective fiction) what our present would be like if the world had voluntarily gone into the stationary state some 150 years ago? To be sure, as Mill was the first to insist, and as it will be increasingly important for us to remember, what he meant by that expression (in our terms, a zero degree of economic growth) is not the absence of change, or the end of progress either technological or spiritual, that is to say in the *quality* of life. But it would have meant, especially during the period of explosive material consumption and production that corresponds with our historical fiction, a much lesser quantity of goods than were actually enjoyed (and, periodically, massively destroyed) by the many people in the world who were their disproportionate beneficiaries. No matter how we measured such a virtual "loss" (and thanks to our ever deeper statistical memories and the wondrous tools of modern economic analysis, it could be done within a meaningful range), and even allowing for all the countervailing benefits of a better custody of the natural world, and even of human beings themselves, most of our contemporaries would find the missed opportunity intolerable – no less than a crime against the potential of humanity.

But if this is so evident for the past 150 years (or 200, or however far we want to go back – but this last period was immensely more productive than any previous one), then why should it be different for the *next* 150 years, or even only for the next 100, for the next 50 years, especially in those parts of the world that have not yet reaped the benefits of all the growth in the past?

For even if the question is badly posed in this way, it is the form in which it has to be answered by those who would now advocate for something like a stationary state. So successful has the model of economic growth proved to be, the "progressive state" initiated at the end of the eighteenth century, in spite of the profound crises that have periodically set it back and of the human cataclysms that have marked its history, that it has become a norm, a near universal horizon of expectations. Not only do those who have profited the most from it assume that their standard of living should remain at least as high, while those who have not yet shared in its benefits justifiably aspire to reach the same level of prosperity, but nearly everyone unquestionably expects that it should keep increasing. Such an expectation has indeed become so naturalized, so obligatory, that no ambitious politician of any persuasion (including, so far, all but the most marginal of the Green partisans) would dare to contradict it. This assumption seems all the more unassailable because of the vast and compelling body of economic theory that very largely sustains it (and there is nothing surprising in this, since as we have seen political economy has from its inception identified its goal as the promotion of growth). But even setting aside the political and economic arguments themselves, it is only to state the obvious to observe that for nearly everyone in the modern world, the progressive state is by now the normal state; anything less is perceived as an anomaly, a crisis, a failure to continue on the natural course of human progress, which is to keep growing.

The burden of proof is therefore on the side of limits, on those who believe that indefinite growth is no longer possible – an inversion of the situation that prevailed when the founders of liberal economics dared to question the traditional, agrarian and nature-based notion of a finite quantity of wealth, and theorized the possibility of the progressive state. When their challenge was to show how production and exchange could be freed from the limits that had from time immemorial been those of natural cycles and local conditions (what we now call ecological limits), ours is to demonstrate why the process of liberation they initiated, and that has proved so successful for so long, should now be reaching its limits. And it is good that we should be so challenged, for how can anyone be sure? It is now widely understood that one of the major factors of economic growth, and the one that has become increasingly determinant in advanced societies, is innovation, invention, the scientific discovery of new and ever more productive technologies and use of resources, often wholly unforeseen by previous generations. This explorative and creative process is the glory of humanity and the most astonishing power of modernity, and barring a complete collapse of civilization, there is no reason why it should come to an end in the foreseeable future. What is almost guaranteed, to the contrary, is that it will keep feeding future growth, in ways that cannot possibly be foreseen.

But technological innovation is not the only factor of growth. Now just as much as in the past, economic expansion must also be "fueled" by material resources, and indeed by fuel itself. Even in the most advanced societies, the new advances in "software," as we often analogically think of the immaterial side of any productive process, depend for their implementation on vast amounts of very real hardware and energy to run on – as in fact the very software programs that constitute the purely mental factor of the information revolution currently under way rely on warehouse-size data centers and Internet exchanges (albeit by design almost invisible to most us) that consume old-fashioned electricity, lots of it.[1] And so on for almost any of the apparently virtual innovations we can think of. Technological and scientific progress can, and to a certain extent does already serve to attenuate resource depletion and alleviate environmental damage, for example by allowing us to satisfy a fraction of our energy needs with lesser reliance on fossil materials and fewer negative consequences, and it is to be hoped that when we are finally ready, or forced, to face the challenge of limits, research and imagination will work wonders to lighten our ecological footsteps. But so far, notwithstanding the claims and efforts of the greening industry, the overwhelming impact of technological progress has been, as it always was, to increase growth. Not only economic growth, but material growth itself, the continual increase of extraction and consumption of the goods of the earth. The only green revolution we have really known, so far, is the ongoing agricultural revolution started in the twentieth century, fueled by mechanization, chemistry, and more and more oil. It has made it possible to feed many more people, and animals, than ever before, and to vastly better the condition of the former, if not the latter. But it is an ecological disaster.

To come to terms with what really matters, the relation between economic growth and ecological impact, and therefore the question of limits, we need to confront not only the role and the myths of technological progress, but more fundamentally the related illusion of modern economic theory that might be called the fallacy of sectoral progression.[2]

Sectoral progression is real.[3] As they develop and evolve through time, national or regional economies can be observed to progress from a preponderance or even complete dominance of the agricultural and mining sector (extraction of raw materials) to a growing importance of the manufacturing or industrial sector (more or less heavy), and finally to the more advanced mix in which the sector of service activities becomes larger and larger as a proportion of the economy, sometimes to the point of overwhelming domination. This is a relative progression, where more advanced sectors become proportionally more important in the economy, but not necessarily exclusive of the others, agricultural and industrial production remaining present in varying degrees. In many of the most advanced and service-oriented countries in the world – the United States and some European Union members come to mind – the agricultural sector is still surprisingly important for its sociological, even political weight, and heavily protected and subsidized, even if only a tiny proportion of the population remains directly engaged in it, and its actual economic contribution is even smaller.[4] As for the industrial sector, it is still crucial to the economic success of the most admired models of global competitiveness, even if it is well known that many advanced nations are increasingly alarmed, and justifiably so, at the looming disappearance of their manufacturing sector.

It is also true that many if not most of the activities involved in the service sector, even the so-called service "industries," are as modes of production and consumption less materially dependent on scarce resources and destructive of the environment than those of the industrial and even the agricultural sector.[5] Algorithmic financial traders, designers of software programs, or even less flamboyant insurance agents, all of whom can generate spectacular amounts of wealth in a modern service economy, only really need intellectual creativity and a good computer or two, sometimes not even an air-conditioned office, or an office at all – although some make heavy use of the aforementioned datacenters. Counter-examples can certainly be found, such as race-car drivers, or for that matter truck drivers and airline pilots, and many more humble cases in between, such as dry-cleaners or skiing instructors (gravity is renewable, artificial snow is not), but as an average we would indeed find that the environmental demands and impact of the service activities themselves are less heavy than those of industrial production and contemporary intensive – and equally industrial – agriculture.

Are we to think, then, that as more and more countries and parts of the world move up the sectoral ladder, we are globally becoming less heavy on the earth? Is sectoral progression, by its very nature, a greening progress that will save us from confronting limits? This is indeed the impression that, with a certain logic and persuasiveness, is often conveyed by the concept of sectoral progression. It is inherently imagined to be ecologically progressive. But what happens to those heavily taxing and polluting activities characteristic of the primary and secondary sectors as the economy progresses to the service sector? Are we to understand – this is of course where the fallacy lies – that they simply disappear?

To begin to answer this question, there is no better way than to go back once more to the Four Stage Theory, and more precisely to an aspect of stadial progression that has so far been left unexamined, and may yet be crucial to the very concept of economic stages, indeed of any stadial theory whatsoever: what is the relation of succeeding stages to those that preceded them? Or to echo more closely our formulation in terms of sectoral progression: what happens to earlier stages when they are superseded, or replaced, by a more advanced one; do they just vanish into obsolescence?

Of course, stages (the economic stages of the classic stadial progression) are not the same thing as sectors (the economic sectors of modern sectoral progression). For one thing, as we have already recognized, all three economic sectors may and normally do coexist at the same time in a given economy; it is only in their relative importance that they succeed one another in sectoral progression. Stages, on the other hand, do supersede and wholly replace one another through stadial progression. Yet it is not hard to recognize a strong family resemblance, indeed a sort of kinship relation between them: as Vivienne Brown has already recognized,[6] the economic sectors of contemporary societies are in a very real sense the descendants of the classical stages, even if not all the latter have survived, and some have been visibly transformed. There are practically no hunters and gatherers left on earth, and certainly no social group that could be said to survive wholly at the first stage of "the chase." For some coastal people, the kind of small craft traditional fishing that can be associated with the stage of "hunters and fishers" still provides a substantial part of their livelihood, if increasingly precariously, but even they are integrated in economic systems that rely on more advanced sectors of production. Most of the actual fishing taking place in the world today bears no resemblance to first-stage gathering, and even that industrial-scale seafood take from the open seas has now been overtaken by captive fish cultivation that is more akin to mass animal breeding than to any form of fishing. So much for the first stage.

Almost the same can be said for the second stage of the early economists, the age of shepherds. There remain in parts of the world some traditional pastoralists whose semi-nomadic mode of subsistence resembles the models familiar to eighteenth-century stadial theorists, but no national economy that is still predominantly at the pastoralist stage. The vast majority of animal husbandry practiced in the modern world is at least at the third stage, if not at the fourth one. What this means is that the first two stages of the classical theory are now history; the whole contemporary globe is at least at what was then conceptualized as the third stage, and beyond. And so if we set our stadial clocks forward to the present, we see that the primary sector does in fact correspond to the first of the classical stages still extant, agriculture, to which we must now add a use of land that was already included in the original conception, but has proliferated exponentially since then, the extraction of mineral resources, not the least of which are fossil fuels.

It is true that in addition to the disappearance of the first two stages, the comparison between the stage theory and sectoral progression also presents us with the case of a (retrospectively) missing stage in the original formulation – a stage whose conspicuous absence we had already noted when first expositing the classical theory in Chapter 2. Unlike their immediate successors of the nineteenth century, and of course ourselves with the benefit of hindsight, the eighteenth-century progenitors of the theory of economic stages had either not foreseen, or failed to sufficiently recognize the looming importance of the industrial revolution whose stirrings they were clearly witnessing, and to identify it as the next stage. There is no doubt, however, that according to their own logic of a social progression through modes of subsistence, later known as modes of production, the most important developmental step in the history of humanity after agriculture has been what we now recognize as the (still unfolding) industrial stage – as does the nomenclature of sectoral progression by identifying the secondary sector as industrial. In order to reconcile the two systems, we must therefore insert within the Four Stage Theory the industrial stage missing from the original version. But what the founding fathers of the theory may have missed by lack of foresight in the case of industry is more than compensated by their prescient choice for the last stage: commerce. As we had also remarked when first encountering the term, commerce is not really a mode of production. But neither is "service," the umbrella term chosen for the tertiary sector in the modern sectoral progression. The terms are not the same, and meant to emphasize different aspects of a commercial or service economy. The classicists, as is well known, were advocating the benefits of free trade for expanding the wealth of nations; the moderns want to distinguish service activities (all of which are of course commercial) from strictly productive (primary and secondary) ones. But in their suitable vagueness, the overlap between the two terms is considerable, and if we think of the proliferation of the financial sector in latter-day service economies, we may well consider them nearly interchangeable. The final stage/sector is in both cases conceived as that of commercial service.

Thus it is not merely an analogy to see in the model of sectoral progression a modern incarnation of the stage theory. Sectoral progression does in its own way repeat in contemporary terms, however inexactly, the passage through the generic stages outlined by the classic theory. If we were looking for an analogy, admittedly a daring one, we might say that economic sectoral progression within each modern country repeats the universal stadial model of social evolution in the way ontogeny recapitulates at the individual level the phylogenic evolution of living species. Without insisting too seriously on a parallel both geneticists and economists would have reason to repudiate,[7] let us say that at a minimum the concept of sectoral progression reveals the persistence of stadial thinking in contemporary economic theory, with all its illuminating power, as well as its blind spots. This is at least enough justification to ask about both the fundamental question posed, and somehow hidden, by the very idea of stadial succession: what happens to the stages, or sectors, that are supposedly left behind?

Let us go back to the mother of them all. Even though it is no longer directly relevant to sectoral progression, the transition from the first stage is so paradigmatic, and revealing, that we cannot simply skip it. The passage from hunting and gathering, or foraging, to agriculture, by now one of the most thoroughly studied by anthropologists, archeologists, prehistorians, and many others, was much more complicated, contentious, and lasted much longer than eighteenth-century theorists thought.[8] They did not see it even when it was in front of their eyes in their favorite (but mostly wrong) example of the first stage, Native Americans. And indeed for millennia it would not always have been easy to tell where collecting ended and cultivation began, and both continued to be practiced coevally. But now it is over. As we noted above, there are practically no humans left to live by hunting and gathering alone in the world, and so we can see what happened to the mode of production on which the first stage was based. Some people of course still collect – indeed most people may well have occasionally collected some wild species – berries, mushrooms, or seashells and crabs. Some people continue to be avidly devoted to hunting and fishing. Indeed, it is well known that in feudal societies, long after agricultural settlement and reliance on domesticated animals, hunting had become an aristocratic privilege, notwithstanding the stubborn persistence of poaching. Fishing always was more humble, if no less beloved. But the pattern is clear: the activities that provided the subsistence of the first stage are now considered leisure. They may well still make a difference in the lives of many marginal peoples, but in developed countries, they have become economically insignificant.

Does this mean, however, that modern humans eat fewer plants and animals? The question can only appear rhetorical, and most likely just silly. Of course we don't. Many more of us eat much more of both. The passage to agriculture allowed our ancestors to multiply greatly their food production and to feed many more of their own, which is why they invariably overwhelmed, conquered or displaced those who had not adopted it, and the reason why they developed it in the first place. Agriculture gradually but spectacularly increases the size and yield of natural vegetable and animal production by cultivating or domesticating it. That is what agriculture does. Through the use of selection and intentional crossbreeding, and of progressively more efficient techniques from slash and burn to fallow alternation to manuring and then scientific soil fertilization and finally direct genetic code manipulation (but we are then well past the third stage), it has made it possible to grow vastly greater quantities of much larger edible plants in the same cultivated area. Through domestication, selective breeding, and closer to us artificial feeding, medicalization, and more genetics, it has also transformed nimble fast runners (for good reason, all our terrestrial domesticated animals except for dogs and cats were originally prey rather than predators) into huge producers of meat or milk or eggs on various numbers of legs (sometimes barely standing). The only reason to insist on such well-known facts is to make the pattern of stadial progression clear. By moving from hunting and gathering to agriculture, human beings did leave behind the mode of subsistence of the previous stage and rendered it obsolete, but they did not leave behind the subsistence basis of the earlier stage. On the contrary, they multiplied it by transforming it into the mode of the next stage. In the case of agriculture, by cultivating and domesticating it. Later stages do not replace the material basis on which previous ones relied, they turn it into their own mode of production and make it grow.[9] They multiply growth, always. That is the point of stadial progress.

It should nevertheless be remembered, before the decisive break opened up at the next stage, that with the domestication of animals, the populations of Eurasia did not only increase the quantity of meat they could consume. They also acquired (in addition to the common diseases that would later cause such devastation among American populations on first contact) a life-changing means of transportation and energy: animal power. This transformed not only agriculture by allowing deep plowing, and the speed of travel and communication, and the art of war, and the fate of the bison once the horse was reintroduced in North America, but pervaded every aspect of life. In the developed world, it is easy to forget just how recently not only all the countryside, but all the cities were filled with the draft and work horses on which so much movement relied. Animals were not the only source of power. From late Antiquity, wind became increasingly important as a source of energy both at sea and on land, and the use of waterpower can without exaggeration be credited for the beginnings of industrialization. Animals, of course, unlike the two sources of energy just mentioned, were themselves dependent on a lot of energy in the form of feed provided by agriculture. Yet practically all the resources on which the third stage was based were potentially renewable. This does not mean that agricultural societies have historically been immune to what Joseph Tainter and Jared Diamond called collapse. There have been clear cases on all continents of cultures entirely based on agriculture that eventually drove their resource base to extinction through deforestation, soil depletion, or the salinization effects of long-term irrigation, for example. Still, as many contrary examples attest, particularly in the Old World, the mode of subsistence that defines the third stage: traditional, animal-based agriculture, was able to achieve what we now call sustainability. It was the second time humanity found such a possibly self-perpetuating equilibrium between human population and the limits of natural resources, the first and far longer being of course the first stage itself, the multimillennial period of hunting and gathering.

Whether the fourth stage, the one we are still living in, or a next stage as yet unreached, will ever again be able to attain such a viable equilibrium state on the scale of the whole globe is eventually the biggest question we must seek to answer. But for now, let us continue with a more limited subset of it, the relation between our present stage and the preceding one, agriculture.

For the eighteenth-century theorists who conceived the fourth stage as Commerce, the transforming world in which they lived was still largely agricultural. Indeed, as they were observing it as beneficiaries of what we may truly call the first globalization, toward the end of the first wave of European colonialism, one of most visible transformations and sources of new wealth was that within the previous 300 years all the main plants and agricultural products of the New World had been transplanted to the Old, and most of those of the Old World, together with all domesticated animals, had been made to colonize the New. Even the most atrocious human aspect of this first wave of globalization and migration, the Atlantic slave trade, was as is well known directly related to the highly profitable plantation economy. Although they were variously aware of the looming importance of new technologies and manufacturing – Adam Smith's famous emphasis on the division of labor was according to himself chiefly applicable to factory production – they were not envisioning the replacement of agriculture as the economic base of society. What they wanted to abolish was the protectionist system of duties, bounties, commercial privileges and monopolistic rights of exploitation (such as those of the colonial companies) so as to liberate trade – that is the meaning of liberalism – by expanding the market for all products, including agricultural. As Commerce, the fourth stage was not meant to supersede agriculture, but to increase it by turning it in its own mode, that is to say, by commercializing it. Such is the repeated pattern and logic of stadial growth: commerce commercialized the agricultural basis of the previous stage, thereby making it bigger, a process of increasing commodification we still observe in our own time when hogsheads are (virtually) traded on the Chicago exchange, and uncountable numbers of as yet unseeded grains are exchanged in the global futures market.

Of course, without diminishing the importance of this process, and with the benefit of hindsight, we now recognize the fourth stage as having been even more markedly industrial, and industry is obviously about a lot more than agriculture. It is about mechanization, certainly, although sophisticated machinery such as cranes and pumps and mills of various kinds had been developed through time and widely diffused before industrialism. Nevertheless, without offense to the investigative genius and pragmatic discoveries of previous ages and cultures, industrialization is undoubtedly the outcome and practical result of the extraordinary revolution in fundamental science started in the seventeenth century, and continued at an incremental pace ever since.[10] Notwithstanding the increasing technicality and artistry of the use of metal ores ever since the Bronze Age, modern industry represents a huge quantitative jump in the exploitation of minerals, and an equally qualitative one in the development of chemistry. And finally, it can never be said enough, it is the first mode of production in the history of humanity to be primarily based for its energy needs on the mass of bio-power we now call fossil fuels, slowly accumulated since the origins of life on earth. Whatever qualifications or additions one might want to propose to those different aspects or causes of industrialism, nobody would dispute that it qualifies as a world historical stage, unlike any that preceded it, and in terms of its effects on economic growth, incommensurable. And yet even defined as Industry the fourth stage did not do away with agriculture, nor did it diminish its scale. Instead, following our familiar pattern, it turned it in its own mode, i.e. it industrialized it, making it in the process vastly more productive.

This process (like the previous transition) was relatively slow at first, in part because the first industrial machinery, powered by the steam engine, was not easily adaptable to agricultural use, and also because as we just recalled, there are different aspects to industrialism, some of which, like organic chemistry, and *a fortiori* genetic engineering, developed later than others. As late as the middle of the twentieth century, much of agriculture even in developed countries was, slightly incredibly, still animal-powered, fairly free of chemical pesticides, and in no negligible degree animal-fertilized – it was, we could say with some nostalgia, still largely organic. No longer. As a result of mechanical improvements, progress in soil chemistry, the development of fertilizers, and above all, the oil that made it all possible, the transformative process culminating in what became known as the Green Revolution has thoroughly industrialized agriculture, making it possible to produce enough food to sustain more than adequately previously unimaginable numbers of people, and in this sense, proving the early Malthusians wrong. The fact that such an industrialized agriculture is no longer sustainable, and not only because of its dependence on petroleum and gas, does not in any way diminish the logic and efficacy of the stadial progression it so perfectly illustrates.[11] Fourth stage agriculture is that much more productive than its antecedent precisely because of its industrialization, which means to a large degree its being boosted by fossil fuels and their derivatives. Indeed, it would not have been industrialized, and as it were fossilized, if that had not enabled it to grow more, in both meanings of the word. Stadial progression, at whatever stage, is a qualitative process that always results in quantitative increase. Progress, in this sense, always means growth.

Because we are still very much at the fourth stage, whatever the heralds of post-industrialism may claim, it is necessarily speculative to predict that the same pattern will be repeated if we move to another stage. But to the extent that, in a more limited way, we can be considered to have already entered into recognizably new sub-stages of industrialism, some confirmation of the trend may be observable. The most credible candidate for such a new technological stage in the last few decades is what is usually referred to as the Age of Information, or the digital revolution, and there is no doubt that information technologies are having profoundly transformative effects in many if not all areas of contemporary life, including modes of knowledge, communication, and invention, with future prospects as yet unimaginable. But there is no sign at all, nor even any likelihood, that information will replace industrial production or diminish its size. Instead, following the usual pattern, what we see is that information is informatizing industry, or digitalizing it, thereby making it more efficient, precise, autonomous (from human labor), and productive. For all its alleged virtuality, it is having hugely accelerating effects on material production and consumption.[12] And without insisting again on the very real materiality and energetic greed of data centers and exchanges, let us not forget that the marvelous electronic devices that put the information revolution on everyone's desk and in everyone's hands – inextricable amalgamations of plastics and rare metals – have the fastest obsolescence (replacement rates) of any human artifacts, except perhaps tooth-brushes (the latest electric incarnation of which do in fact contain micro-chips).[13] The information age is not just producing more information, it is producing more material goods.

Of other possible candidates for technological stagehood, genetic engineering, the recent and still very young offspring of DNA science – potentially more profound in the long history of humanity for its shattering anthropological and ethical implications than information technology, which had at least a worthy precedent in the invention of writing – has so far not had time to have enough effects beyond medicine and agronomy for us to know how it might alter our relations with other species, even indeed our notion of species, with the possibility of creating and multiplying as yet inexistent life-forms for our own consumption. It too, besides its potential for making unprecedented qualitative changes in our existence, promises undreamed of opportunities for growth in both senses of the word. Whether these technological advances, or any other yet to come will ever be viewed as decisive stages in our historical progress is of course impossible to tell, but what we can expect from the past record and the logic of stadial progression is that, barring an epochal paradigm shift in human priorities, whatever emerges as the next stage will inevitably turn the productive base of the previous one into its own mode, and make it even more productive. Meanwhile, without having to wait for a distant future, we do know that somewhere in the American Mid-West, an armada of GPS guided tractors is sowing the genetically modified seeds of next year's harvest: an informatized, geneticized and fossilized agricultural production already sold on the world futures markets as bigger, as well as better than ever.

Sectoral progression works somewhat differently. It too assumes a normative succession through time, but as we have seen what it measures is the shift in the relative importance of the three economic sectors compared to one another. According to this generally well confirmed historical pattern, the agricultural and mining sector is first predominant, then an industrial sector gradually develops and in turn become preponderant, and finally a third sector, that of services, gains a larger and larger share of the economy in relation to the other two, so that advanced societies are expected to end up dominated by the service sector. It is not exactly a question of successive sectors replacing one another, as with stages, nor of transforming previous ones, although everything we saw about the retroactive technological effects of stadial progress still applies here. Of course, some economies can still be almost totally agricultural, and at least in theory one could imagine the case of a society from which agriculture has been completely eliminated and that devotes itself entirely to industrial production (necessarily in exchange with other more diversified economies). And there are indeed a few examples of small nations, such as Luxembourg, that are exclusively service oriented. But that is not the norm, and as we shall see it never could be so. What is normal is for modernizing countries to progressively emphasize the three successive sectors in their relative proportion to one another: the more advanced the economy, the more predominant the later sectors will eventually become. The question is, what does this tell us not about their relative, but about their absolute size?

For the first two sectors, much can be learned from stadial progression. In the most advanced countries, the agricultural sector can seem very small in comparison with the others if we consider solely the number of people who are directly involved in it, now approaching a few percent of the population.[14] But that is precisely because modern agriculture has become so industrialized, and productive. In those same countries where so few people actually work on the land, the agricultural sector still makes substantial demands on the supporting industries that make such productivity possible: mechanical, chemical, pharmaceutical, and not least (fossil) energy, not to mention its direct effects on the water supply,[15] both on the intake side for its enormous irrigation needs, and on the outflow for its contamination of soils and aquifers. It is also responsible for Global Warming gas emissions vastly disproportionate to its share of economic activity.[16] Such an agricultural sector is as heavy on the earth as industry itself, precisely because it is industrial.

Still, it is for good reasons that we normally think of the second sector, industry itself, especially in its classic nineteenth and twentieth century "heavy" manifestation, as presenting the central dilemma for environmental concerns. If it has made it possible for a larger and larger proportion of an exploding human population to enjoy so many material goods and conveniences, an unprecedented and continuing rise in the global standard of living whose benefits should never be underestimated, this has been achieved at the unbearable and unsustainable cost of irreversible planetary depletion. From an ecological point of view, what industry does is to accelerate what is known as throughput, the process by which useful and often limited material resources are turned (while producing a good of more or less lasting value for us, but almost never everlasting) into permanently useless and often detrimental by-products: the transformation of sources into sinks. In terms of entropy, it is the acceleration of the entropic process itself: the one-directional degradation of low entropy (organization) into high entropy (disorder).[17] Historically, each technological innovation has had the effect of boosting this process, even as it was opening new productive capabilities, or transforming old ones, not only because increasing production in its specific domain or inventing new products is the *raison d'être* of innovation, but also by what is known as the multiplier effect: any increase of activity in a particular area of the economy – and especially in the industrial sector – stimulates economic activity in other areas. Economic growth – eventually any growth – is entropic.

It is true that during the last few decades, due to the diminishing availability (and therefore rising costs) of material and energy sources, and to a new consciousness of the harmful consequences of industrial pollution and environmental destruction, science and technology have begun in earnest to pursue within the industrial sector itself the twin goals of efficiency and environmental preservation. Since this search for ways of satisfying human needs that are less taxing for natural resources and damaging for the ecological balance of the earth is an indispensable step in the transition toward living within limits, there is every reason to rejoice in this development. It could signal a real turning point in the direction human progress has always taken if – and it is a serious if – the aim is not merely to keep consuming more while mitigating the negative impacts of continuing growth. In the richest parts of the world, many highly developed countries have thus since the last decade of the twentieth century been able to curb industrial pollution, improve sometimes dramatically the quality of their air and water, and get rid of their worst sinks, in part by adopting more efficient and cleaner fuel burning and alternative energy technologies. The lower emission standards achieved in the automobile industry are a good example, resulting in reduced CO_2 levels even while the number of motor vehicles was increasing – seemingly proving that growth is at least in some areas compatible with environmental improvement. Except for a well-known obfuscation whose full significance we will contemplate when we next look at the tertiary sector: during the same period of time, the same countries have massively outsourced their dirtiest production, and even outsunk their own garbage, to poorer parts of the world.[18]

By contrast with industry, it makes intuitive as well as some real sense that services – the tertiary sector – like commerce with which it is intimately connected, should appear much more abstract, almost immaterial in its impact on the world, approaching the idea and the ideal of a non-entropic process. And indeed, as we had already noted, some forms of services, sometimes among the most valuable in economic terms (art, software design, legal services, insurance, even the infamous financial sector), do not in themselves require much in the form of material resources, and produce relatively little environmental nuisance. This is assuredly not true of all services, as can be ascertained by counter-examples such as the hotel and tourist industry, transportation, or the ever-expanding medical sector.[19] Still, on balance, service activities in and of themselves generate less material throughput, and have a lower environmental impact, than industrial production. It follows that as an economy evolves through normal sectoral progression from a domination of industrial activities to a larger share devoted to services, it should become lighter on the earth. As the relative importance of the industrial sector diminishes, so will the material demands on natural resources and damages to the environment. Sectoral progression is then thought to have a virtuous effect. The more a society develops its tertiary sector as it becomes more affluent, the less its environmental impact will be, relative to its overall economic activity. Ecological footprinting ought to be inversely proportional to sectoral progression.

Economists like to express this virtuous trend by what is known as a Kuznets curve. Simon Kuznets originally used the curve that bears his name – an inverted U or bell-shaped curve – to illustrate the correlation of social inequality with economic development,[20] but it has become widely used to chart the evolution of various environmental impacts in relation to the progression of the economy. And so in graphic form, we would see that as a society begins to develop, its use of natural resources and environmental degradation first rise steeply – as indeed they do with industrialization – but then as it further progresses to what we can now identify as the tertiary sector, its material impact should gradually level off, and will eventually start and continue to fall, perhaps indefinitely, along the descending side of the graph.[21] Kuznets curves are hopeful curves. And as a matter of fact we can point to some noted empirical cases in conformity with this model. In the most developed countries, the last decades of the twentieth century have seen a real decline of water and air pollution (except for CO_2) in spite of continued economic growth. Even for the all-important carbon dioxide itself, whose emissions have continued to increase, it is nevertheless significant that its relative rate of growth has been measurably diminished thanks to improved fuel efficiency in motor vehicles, and ameliorations in energy intensity generally, while the number and usage of cars and trucks on the roads was rising unabated.[22] These are indeed hopeful trends, since they provide fairly strong evidence of the alleviating gains that can be achieved by technology, when put to the right use.[23]

But the partial truth of these relative decreases in environmental degradation in the richest parts of the world risks hiding the overwhelming fact of the absolute growth of material consumption and ecological footprint in all the economies that have achieved a highly developed service sector – hence the ultimate fallacy of sectoral progression. For relative as opposed to absolute is a crucial distinction here. Advanced economies, as they progress from the secondary to the tertiary sector (in relative terms), become relatively less active in industrial production and more in services – that's what sectoral progression means – and therefore, as producers, relatively less involved in the most resource-hungry and environmentally damaging activities, but this does not mean that, in absolute terms, they become less reliant on industrial products and more thrifty in their energy use. On the contrary, they hold the absolute record in these categories. After all, there is no dispute that the countries with the most highly developed service sector also have the highest standard of living, including the highest standard of material consumption, which means the biggest ecological footprint. And far from beginning its hoped-for descent, the curve only keeps rising.[24] This should be enough to dispel the fallacy, or at least the excessive hopes placed in the transition to a service economy, even if it remained a baffling paradox. But there is no mystery. As is already evident from the above distinction between relative and absolute measures of material economic activity, the fallacy rests on a failure to make another, equally crucial distinction in a service economy, that between production and consumption, even though people, as economic agents, are of course both producers and consumers.

The software designers, corporate lawyers, and financial wizards who are, we remember, the heroes of the service sector for creating so much wealth with so few material resources, are nevertheless not exactly known for living lightly on the earth. As producers, they may be relatively harmless, if not altogether virtuous in terms of environmental impact, but as consumers, they can be an absolute menace. And so, to a lesser extent, are even the most otherwise frugal of frequently flying academics. So are the great majority of those of us who live by the average standards of the developed world. As for the least privileged members of the affluent economies (though even they, by comparison with most inhabitants of the underdeveloped world, would be considered ogres), when they are not farm laborers or low-qualification workers in the remaining industrial sectors, they are massively employed in the retail and consumer sectors, as salespersons in small and large stores, as checkers in supermarkets, suppliers of countless outlet malls (accessible only by car), or eager sales assistants on autorow. Though they may not be able to buy a lot themselves, they sell things to those who can. And it should be no surprise. A service economy is necessarily a consumer economy, and consumers consume goods, sometimes immaterial, or almost so, but still in most cases very concrete indeed, or plastic, or fabric, or metallic: matter and embodied energy.

Service economies have not become less dependent on material things – or only relatively, marginally less so. Instead they now rely on material things that are produced elsewhere, outsourced and offshored to other parts of the world whose economies are lower on the sectoral pole. This means that if we counted in all the resource use and pollution that has thus been expatriated (a literal or geographical form of what economists call externalization), if we internalized the ultimate sources and sinks of service economies by repatriating them, their footprint would look even worse. In the real global world, the environmental Kuznets curve has woefully failed to materialize.

And this is the place to recognize the parallel argument proposed by Brian Czech under the more immediately understandable label of "Fallacy of Self-Sufficient Services," first adumbrated in an earlier book and then fully developed in *Supply Shock* (Czech, 2000; 2013).[25] What we have been exposing through the historical development of stadial and sectoral succession, Brian Czech demonstrates through the analogy – or similarity – of the hierarchy of economic sectors with the biological model of trophic levels.[26] In the economy of nature, as he aptly calls it, "[t]rophism refers to the transfer of energy and nutrition from one organism to another in the process of feeding" (Czech, 2013, 173). The first trophic level, "primary production," is that of plants, which are the only organisms capable of producing their own food through interaction with the soil and the sun. On the next trophic level (that of consumers) are the animals (and us) who eat plants, and on the top level (secondary consumers) are the animals (and us) who eat animals who eat plants. On the highest level, adds Czech, animals "such as lions, eagles, crocodiles and sharks, are sometimes called 'super-carnivores'"[27] (ibid., 174). The most important fact in the economy of nature is that "no trophic level may consist of more biomass than the underlying trophic level" (ibid., 180). In fact, it must contain less usable energy. This is the direct consequence of the second law of thermodynamics, the law of entropy. In the natural world,

> each trophic level contains approximately ten percent of the biomass of the next-lower trophic level. For example, in an ecosystem with 10,000 tons of

> producers [i.e. grass], one may expect approximately 1,000 tons of primary consumers [i.e. elk] and 100 tons of secondary consumers [i.e. mountain lions].
>
> (Ibid., 181)

This results in a hierarchical, pyramidal structure of trophic levels, with each one necessarily smaller (in the example above, much smaller), that the one on which it rests.

This is the economy of nature. Of course, the human economy is part of the natural economy, which is why we humans can fit so easily in the upper levels of the pyramid described above. But we do more than just act as other animal-consumers. We also accelerate the entropic process at all levels by our agricultural and industrial and service activities.[28] And so Brian Czech draws another, identical pyramid for the human economy. In this one the bottom level is appropriately that of agriculture and resource extraction, the primary sector, from which all other economic activities ultimately depend. It is striking, and this is true of ecological economists in general, that he thus recognizes the primacy of Land, the factor that has become so neglected by modern growth theorists, or at best included in Capital (at least producing the important concept of natural capital). In this recognition of the unique function of Land, ecological economists go back, sometimes unknowingly, but quite consciously in this case, to the fundamental principle of Quesnay and the Physiocrats in the eighteenth century, who insisted that all economic value derives from agricultural production – to which we now add the extraction of mineral resources, also provided by the earth. Brian Czech calls this principle, and his entire Chapter 7, "The Trophic Theory of Money"[29] (Czech, 2013, 171–193).

Now for the parallel we are pursuing with sectoral progression, it is true that in his pyramidal representation of the human economy, Czech does not distinguish among the higher trophic levels between industry and services.[30] Instead the equivalent function is indicated on each side of the pyramid by "Service" and "Providers" (ibid., 192). His mid and top trophic levels are respectively labeled Heavy Manufacturing (a fairly close approximation of Industry), and Light Manufacturing. But the fundamental principle remains: whether they are designated as manufacturing or services, the matter and energy demands of the economic activities at each ascending trophic level cannot be greater than can be supported by the primary sector.[31] And as the pyramidal shape of the model indicates, each has to be smaller than the lower one. It is also true than in the case of the human economy, we cannot know in advance the fixed proportion by which the bio-mass or total embodied energy at each trophic level must be reduced (as in the "approximately ten percent" of the economy of nature). This entropic loss depends crucially on the energy efficiency and resource use maximization of our industrial and consumer practices, including the extent and efficiency of our recycling. But it remains that the higher trophic levels, or sectors, cannot expand independently of the lower ones. In Czech's terms, services cannot be "self-sufficient."

Back to our terms, this means that sectoral progression does not supersede previous sectors any more that stadial progression replaced earlier stages. On the contrary, so far, as the richer economies have been able to devote a greater share of their wealth to an increasing range of immaterial services, they have also kept increasing their appetite for natural resources and energy, and their effluence into bottomless sinks, even if those are more and more frequently outsourced and outsunk out of sight, because material consumption is still how we measure the good life, tangible wealth itself, not out of thin air, but by the footage of our real estates. Notwithstanding widespread fantasies about post-industrial society, service economies are still very much at the fourth stage, still the primary beneficiaries of its extraordinary productive capacity, and the willing subjects of its originary logic and ongoing purpose, which is that of growth.

Only by now the fourth stage has accomplished what its earliest thinkers, and even those few who later came to contemplate the desirability of achieving a stationary state, had never really dared to imagine: it has not only reached, but overtaken the natural limits of what made it possible, and would have made it, perhaps, endurable. This is a reality even the most environmentally disposed among us still find difficult to accept. The world is so vast, the earth's eco-system apparently so resilient, so forgiving, at least for a long while on the scale of human experience, that it challenges our ordinary understanding. And indeed, the early warnings of such an impending possibility were for a long time greeted with disbelief, if not derision, and still are by some. But in fact the true crossing of sustainable limits, the invisible ecological threshold, was probably reached just around the time of its most famous heralding, the Report for the Club of Rome justly entitled *The Limits to Growth*, and when the less widely recognized but most prescient of pioneering ecological economists, Herman Daly, had already began exposing almost everything we need to know about it.[32]

By now, according to the most sober and well informed measurements of global ecological footprint, we are all of us already using almost one and a half earths, which means of course grievously abusing the only one we have.[33] And if every inhabitant of the planet managed to live as the average of the developed world, which consumes a whole earth all by itself (20 percent of the world population use around 80 percent of all resources), we would need several earths, by most accounts three or four[34] – an extrapolation which begins to seem as tragically absurd as Malthus' impossible projections of population increase. And that is without taking into account our own projections, inevitable for all we know, of how much global population will keep increasing within the next 50 years, in most parts of the world with legitimately rising expectations of a greater share.[35] The prospective numbers, when we think about them, are mind-numbing.

The point of these extrapolations, however, is not to be numbed by numbers, but to persuade ourselves of the need to internalize – in all senses of the word, and ultimately in an economic sense as well – the reality of limits, and the necessity of thinking again in terms of limits. Not any particular limit (although all limits are particular, and require their own thinking), but the new, or returning condition, of limitedness itself.

Over the past several decades, public and official attention has increasingly become focused on one looming limit, the level of concentration in the atmosphere of CO_2 and other gases responsible for the greenhouse effect, hence for global warming. Although some uncertainties may remain about the precise number of parts per million that present a real threat, and many doubts about the actual consequences of current and projected levels, there is no question as to the appropriateness of the overwhelming concern for that limit. Not only because of the potentially apocalyptic consequences of our breaching it further, and at best the dire aftereffects of how much we have already done so, but also because there is a kind of poetic justice about it. The main cause of human-induced global warming, the burning of fossil fuels, is precisely what has made materially possible our extraordinary, and we now understand, unsustainable economic growth in the last two centuries. It is the essence, the veritable fuel of the prosperity we so miraculously achieved, and so it is right, and inevitable, that we should have to focus on it.

Unfortunately, it is also an exceedingly difficult problem to resolve, not only by reason of this utter dependency of industrial economies on fossil fuels, but because greenhouse gas-induced global warming is a question of sinks. As we already noted when first introducing this distinction, whereas questions of limits having to do with sources can at least theoretically be addressed by normal market mechanisms if all true costs are internalized – as could eventually happen with limited energy resources such as oil and gas, though not with the worst CO_2 offender, coal, which is almost infinitely available – overflowing sinks send no conventional economic signals, because they are all externalized; nobody (or eventually, everybody) pays for them. Global warming is thus a global tragedy of the commons. It also poses formidably difficult questions of equity. Between North and South (those who have most offended and those who will likely most suffer, even more so if they cannot offend in their turn), and hence between past, present, and future generations. It is therefore not surprising that there has so far been much talk, and many proposals, but not much action on the scale that is required to come down within safe limits.[36]

But suppose we found a way to respect that crucial limit, say through highly efficient technological means (there are already many such solutions that can make a substantial difference) without changing much else in our way of life. Should this miraculously happen, we would still be – we are still – confronted with other limits, not as frequently recognized perhaps, but just as crucial, if not more pressing.[37] How many more, and how distinguishable from one another, or infinitely multiplying? Jared Diamond identifies 12 "most serious problems" that all result in, or from, breaching a certain natural limit (Diamond, 2011, Chapter 16), but each should really be subdivided into many component parts, in many cases *sui generis* and irreducible to any near kind (is the disappearance of one species of frog less momentous because there are many other amphibians?) Among the most widely recognizable, apart from fossil resource depletions, each with its own distinct schedule, would certainly be the availability of fresh water – the very commodity Adam Smith chose as an example of an essential good without economic value because it was infinite – now being pumped up at great expense from ever lowering aquifers. Then soil suitable for cultivation, also shrinking because of excessive erosion, exhaustion, poisoning or salinization. The same, or the opposite, for the all-important health and wealth of the oceans, and the loss of primary forests and marshlands that are the breathing and filtering organs of the biosphere. And perhaps most tragic of all, if almost invisible, the destruction of countless species and genetic diversity much faster that it can be renewed by evolution. For we are, in the words of Bill McKibben, "running Genesis backward, decreating" (McKibben, 2010, 206). We are engaged in genecide.[38]

All these limits are of course different, and indeed present specific problems with varying degrees of seriousness and urgency. Overfishing the seas and the plasticization of oceanic waters are not the same as the loss of aquatic life in rivers, which is caused by eutrophication from the runoff of agricultural fertilizers, which are themselves overused to compensate for the decline of natural soil fertility. Upper atmospheric ozone depletion from the emission of chlorofluorocarbons is not the same as the greenhouse effect of CO_2 and methane release, coral reefs are not dying from the same death as tropical forests, and endangered polar bears are certainly different from even more endangered tigers, and threatened by different limits. Pragmatically, it therefore makes sense – it is indeed of the utmost necessity – to take the measure of each of these limits separately, and to try to address the specific problems they pose with all the appropriate policies of conservation and applicable technological means at our disposal. It is also eminently reasonable – albeit almost as difficult – to attempt to prioritize them according to their degree of urgency and the extent of the danger they most likely present, both for the present and the foreseeable future. And so there is nothing wrong or misguided about the preeminent focus on global warming which is finally gaining international recognition, and even some semblance of consensus. Given that it is not the most visible limit or perhaps even the most immediate danger, we may even regard the growing consciousness of this particular limit as a sign of wisdom on the part of worldly institutions. As long as it does not screen us from the fact that behind it, indeed in some cases before it, lie many other limits, and that one particular limit always hides, or reveals, another, so that more will inevitably follow. For we are not dealing with one limit that can be solved singly and then everything will be fine, but with an infinite cluster of variously urgent but interrelated limits which all point to one great finitude, that of planet earth itself.

This interrelatedness of all earthly processes in one great organic whole has been studied and theorized in various ways, sometimes beyond scientific knowledge, so much so that it has for many become a veritable belief in the essential unity, even immanent sacredness, of terrestrial life, a kind of geo-mysticism. But we do not have to grant such a metaphysical status to the Gaia hypothesis to recognize that the forces of nature on earth, physical and biological, function as an immensely complex system, both resilient and fragile, in which everything is ultimately interconnected. In such an interdependent world, one particular limit may apparently be pushed for a time without observable consequences, direct or indirect, until a completely unforeseen effect appears in a different area, sometimes on a scale totally out of proportion with the original input – as in the famous butterfly effect demonstrated by chaos theory.[39] Thus some of the predictable consequences of the rise in global temperatures, such as increased desertification and the melting of alpine glaciers and polar ice caps, may be gradual and linear, but past an uncertain tipping point the warming of oceanic waters could have the catastrophic effect of slowing, stopping, or even inverting the course of hemispheric currents such as the Gulf Stream, with immeasurable impacts on some of the most developed and apparently safest parts of the world.

Furthermore, causes and effects affecting different limits may turn out to be linked in feedback loops that precipitate the planetary system into a runaway chain of mutually reinforcing events.[40] The increased desertification and the rise of oceanic waters caused by the greenhouse effect both reduce the earth's capacity to absorb CO_2, which in turn intensifies the greenhouse effect, and the melting of glaciers and polar ice induced by global warming reduces the earth's albedo, the degree to which it reflects the heat of the sun instead of absorbing it, thus further increasing global warming. With so many limits so intimately interconnected all showing signs of being pushed past the breaking point by human activity, the evidence is that we are no longer faced with a few serious problems that can be resolved separately, but with our systemic overwhelming of the earth's balance and renewing capacity, in blatant disregard of the global natural limits that are also ours. To evoke the catastrophic folly of following this course, the famed image of the Titanic hurtling toward its fateful iceberg has often come to mind, but that simile is now hopelessly out of date, as mountains of ice are already melting. If it were not for its inappropriate levity, the more pertinent figure would be that of the cartoon character blithely running in mid air past the edge of the cliff, not knowing that it has already started falling. That is the meaning of overshoot,[41] and it is dead serious.

The only possible conclusion of all this, the truth so really inconvenient that we are so far massively denying it, is that there is a direct, and by now irreconcilable conflict between the logic of growth and the reality of ecological limits. The economic teleology that has served us so well during the last quarter millennium in raising the material standard of living of so many inhabitants of the developed world, and has begun improving the living conditions of many more in developing nations and continents, has also by now become a real and increasing threat to be the natural conditions on which this prosperity depends, undermining the very basis of economic sustainability, let alone progress. Unless we choose to ignore this threat, at our enormous peril because it will not spare us for being ignored, this means that we must redirect our economic priorities away, and downward, from the long-cherished goal of material growth. For the only way to resolve the conflict between growth and limits, a conflict growth cannot win without self-cancellation, is to adjust the level of our material economy to that of environmental limits until they reach again a lasting balance, the condition for sustainability. And because we have now been in overshoot, exceeding our limits, for several decades at least, the achievement of a stationary or steady state economy such as was imagined by John Stuart Mill and called for by many environmental advocates since then, though it should still be our long-term goal, can no longer be sufficient. To redress the imbalance, now and for the near future, we must reduce our ecological footprint, that is to say, on a global scale, diminish the size of the material economy. Decrease the weight of our being on earth.

To be sure, this is not exactly, and perhaps not necessarily the same as what has begun to be designated by the very unlovely term of degrowth,[42] the opposite of standard economic growth. Technically, it is not at all the same thing. On the one hand – the good hand – as we have often recognized, there are many human activities, such as personal services and the development of intellectual property or artistic creation, that undoubtedly produce economic value, and therefore contribute to growth, without drawing heavily on material resources. Moreover, as also repeatedly noted, even when matter and energy are directly involved, the gains in efficiency achieved by technological innovation can significantly reduce the ratio of material input to economic output, thus making it possible to progressively increase the value of goods as measured in GDP – economic growth – without a commensurate increase in material use and depletion, resulting in a lower ecological footprint in relation to the creation of wealth. This has become known as "decoupling," the unhitching of the economic from the material economy. It is obviously a very appealing concept, since it would offer the possibility of resolving the current environmental crisis without having to make fundamental and painful changes in the dominant logic and system on which we depend, without having to renounce economic growth to save the planet. And there are indeed serious thinkers whose recognition of the reality of limits and ecological commitment are not in doubt, such as Paul Hawken,[43] who have long been enthusiastic proponents for such "green growth," even going so far as to see our environmental predicament as an enormous business opportunity. It is to be hoped that they are at least partly right.

For there is another, more sinister hand to the question of decoupling, the side of "bads": first and foremost all those externalities, such as pollution and environmental damages, including the massive greenhouse gas emissions responsible for climate change, that for all their material reality do not register in standard GDP measurements until we try to remedy them, if we can, at great cost. Then, and only then does the negative impact of our footprint begin to count, and, perversely but logically, contribute positively to economic growth. Such is the disconnection – the shadow side of what is optimistically promoted as decoupling – between our material economy and how we measure economic wealth. And that is not taking into account – though some eco-economists have developed impressive estimations – the value of all the irreplaceable resources and environmental riches being destroyed by our economic activities, or the veritably inestimable price of all the species of life we have forever extinguished. It is therefore true that there is much potential for remedial actions – an enormous and widely recognized opportunity would be the thermal insulation of all the terribly wasteful buildings inhabited and used by most of us all over the world – that would be both economically beneficial (contribute positively to GDP) and ecologically beneficial in the long run, in that it would decrease our future demands on the earth (though not without substantial material cost in the present).[44] But this is because of the deeper and more fundamental truth that our ecological deficit, the actual value of everything we already overuse, exhaust, damage and destroy on this same earth is much greater than is reflected in the standard measurement of economic activity. If all our material impact were really internalized, GDP growth would suddenly appear explosive, which our materiel growth truly is.[45] The truth is that, before we can imagine a future when green growth might be possible, and probably for a long time, we need to reduce all growth, economic as well as material, rather than dream of decoupling. However one looks at it, we need to wane.

And that is undoubtedly a very big problem – though it would be better to call it a challenge to give ourselves heart enough to face it. Probably one of the biggest challenges we humans have ever faced collectively, for the reason so evident to everyone that it almost need not be put into words, that degrowth, or *décroissance* in the more natural French original, contradicts the most widespread and ingrained notion of what constitutes the betterment of life, the essence of progress itself. For at least two centuries now, it has become a universally shared expectation, and the predominant measure of societal success, that each generation would have more – more goods, more wealth, more of everything – than the previous one. Although we do not entirely confuse the two in our evaluative judgments, and generally not in our moral decisions, we do intuitively equate better with more in everyday life, and certainly in the economic sphere, that is, for our practical purposes, in the realm of what material goods we can hope to acquire and enjoy. Growth is good, simply and most obviously, because more is better.[46] And as we keep growing more, without necessarily getting better in many important respects, there is yet to be even in the richest societies any widespread questioning of the goodness of more. Growth remains the largely unquestioned horizon of our expectations.

There is one crucial exception to this consensual embrace of growth. Over the last half century, first in the affluent world, and then in most emerging nations and indeed almost all over the world now, people have increasingly decided that having more children is not better than having few.[47] They are no longer condoning the age-old growth imperative to be fruitful and multiply. Of course, this concerns human biological growth itself, not economic growth directly (of which it is nevertheless a fundamental factor). Yet one can hardly deny the profound implications, practical as well as symbolic, of such a decisive reversal of values, and not only because procreation is after all the matrix of all growth. Together with the contemporary movement of emancipation and greater equality for women with which it is closely connected, the equally spectacular decrease in global birthrates beginning near the end of the second millennium is one of the most momentous turning points in human history. Only its spread over a few (just a few) generations and its relative immediate invisibility may still prevent us from being fully aware of its epochal significance. Even if it had no demographic or ecological consequences beyond individual lives, it would have to be recognized as an anthropological and personal transformation of the utmost importance. But of course it does have tremendous consequences for the long-term prospects of humanity on earth. Even though the momentum of life-cycles already started means that global population is still projected to increase to at least nine billion between the middle and end of the twenty-first century – a number that is difficult to imagine as compatible with a world worth living in – the demographic transition currently under way can allow us to project, unless those trends are for some reason reversed, that it will then peak at this level, and perhaps begin to decrease. This is the essential ray of hope, the condition that makes it possible to envisage a sustainable future for humanity. It will one day stop growing, because the size of families is now decreasing. And they are doing so by design, by the choice of women, and men, to favor fewer over more.

This unprecedented event of intentional *décroissance* does not mean, of course, that we can expect it to announce or serve as a model for other choices of less rather than more. Our relationship to our children is so unique that it can hardly be imagined to extend to other realms, and least of all to things themselves. If anything, so far, as a special kind of what is otherwise known as the rebound effect,[48] those people who are having fewer children tend to lavish on each of them such a surplus of material love as to cancel any potential gain and contribute to more growth rather than lessen the familial footprint. Objectively speaking, though obviously not from the point of view of his or her parents, each additional child currently born in the developed world is an ecological disaster.[49] But that is not in itself unique to children. It is on the contrary part of the general pattern of increasing consumption that ought to be reversed.

The larger question then is whether, beyond the exceptional preference for fewer over more so strikingly exercised in the choice of offspring, people in affluent countries might also be inclined, or induced, to reconsider their attachment to material growth itself. It is such a large and difficult question that only elements of an approach can be evoked at this point, but in spite of much apparent ground for pessimism, we should probably refrain from the temptation to despair. Although the actual strength of the current of anti-consumerism and intentional slowing down that is often gathered under the umbrella of "voluntary simplicity"[50] may well seem practically negligible in relation to the still rising tide of global economic activity, a surprisingly large proportion of people randomly surveyed in the rich world – not yet a majority but not very far from half – say they would be willing to do with less material wealth, presumably if other people did so too.[51] This fits well with the findings of various assessments of levels of well-being or of the "happiness index" in relation to monetary income: beyond a certain threshold of comfortable but rather moderate measure of wealth for the First World, further increases do not seem to make any difference, except in perception of relative status.[52] It is comparative rather than absolute affluence that matters most to most people. If so, with enough persuasion that our current rate of material consumption really is a threat to everyone's wellbeing on earth, some consensual acceptance of the need to renounce our attachment to growth might after all be possible.[53]

But even if we could agree to reconsider the value of growth as a moral and philosophical good, and its social inevitability, we would still have to confront the possibly greater technical problem of its vital role within the economy itself. For it isn't just that the modern liberal economic system is designed to produce growth, and rather good at it; it is also dependent on growth to function well, for its own wellbeing, as we express it intuitively when we speak of a healthy economy. Its optimum equilibrium is dynamic, that is to say growing. And most crucially for its social consequences, growth, in the advanced stages of a mature economy such as that of developed countries, is largely driven by gains in productivity. With each progressive innovation in technology or improvement in labor efficiency, more wealth can be produced with the same amount of capital or the same number of people, and the economy grows to the benefit of society. Or alternatively, the same amount of wealth can be produced with less capital (also a benefit), but by employing fewer people – a much more mixed blessing in the best of circumstances. And unless there are consumers to buy the additional products made possible by gains of productivity – mostly other employed people – it is the latter that ensues: people lose their jobs. It is enough to observe what happens in a cyclical downturn, and even more acutely in a recession, to see the social damage caused by a slackening of growth. A fall of just a few points of GDP, say two or three as in the latest crisis, labeled the "great" recession, can lead to at least twice as large a rise in unemployment – millions of people out of work. Is this what the intentional end of growth calls for?

All the serious advocates of *décroissance* must struggle with this question. With various degrees of utopianism, most argue that the only socially acceptable solution entails, indeed necessitates, a reduction and redistribution of labor: each person will work less, perhaps much less, and the accrued leisure time will be shared as much as possible by everyone.[54] It is not surprising that this virtuous scenario, which turns a dire necessity into a beneficial outcome, is embraced by thinkers coming from the political left, since the eventual reduction of labor time has long been a major goal of the socialist tradition, and even famously prophesized by more mainstream figures such as John Stuart Mill and John Maynard Keynes. And surely few would object to an evolution so improbably appealing if it could be achieved without too much disturbance and sense of deprivation. But even if the numbers theoretically add up, and down, the biggest obstacle remains its practical implementation in the real world. Unless an extraordinary degree of cooperation and agreement is reached between countries at very different levels of development – something that is not easy to realistically expect in the near future – global competition could ravage those economies that have intentionally become less productive, and therefore competitive.[55] The alternative path requires an equally unlikely degree of protectionism, if not outright autarchy, to fend off competition. In either case those virtuous countries that attempted to do their fair share to save the global commons in the absence of an environmental contract might be thrown in a downward spiral of economic uncompetitiveness and further loss of employment far below the intended and equitable level.

Perhaps this pessimistic scenario only points out the absolute necessity of international cooperation to achieve ecological equilibrium: *décroissance* will be global or it will not be.[56] It is obviously a political as well as an economic challenge. But there may also be less cause for such pessimism regarding the economic prospects themselves. Even if the advocates of green technology turn out to have been exaggeratedly enthusiastic about its employment potential, it is almost certain that the ecological transition will require a more labor-intensive economy, thus countervailing at least in part the loss of employment opportunity due to reduced growth. Those opportunities, however, may not all be as exalted as the high-tech proponents imagine. A good and in fact supremely important example is that of agriculture. Even if it remains appropriately mechanized and scaled for optimal efficiency, a sustainable, low-petroleum agricultural practice that is respectful of the soil and the environment, wise about its use of fertilizers and water, and less dependent on long-distance transportation – that is to say, largely organic and ecologically mindful – is necessarily more labor intensive than the current industrial model. It requires closer human care, and that is a good thing both for the land and for the growing number of people devoted to nurturing it properly. But even if they earn greater respect and a better living than present farm-laborers do, it is likely to be on a more modest level than bio-engineers and robotic-automation designers. Another example is the already mentioned need for massive building retrofitting and insulation, as well as the conversion of inefficient home-energy and heating – and cooling – systems to renewable self-sufficiency. This presents tremendous opportunities for new employment, but while some of them are indeed in the research and development of composite materials and photovoltaic cells or other high technologies, the greater demand is for the humbler and usually less rewarded skills and labor of the construction trades.

But the most compelling case for the kind of downgrading or down-to-earthing of human activity necessitated by the transition to a no (material) growth economy is probably the turn to generalized re-cycling and reuse. As newly extracted raw materials are increasingly recognized as precious and irreplaceable, and as their cost as well as that of the huge amounts of energy necessary to transform them becomes prohibitive, we should transition from mining and dumping and planned obsolescence to the opposite model: chains of transformation designed from the beginning so that products at the end of their (preferably long) life can be entirely recycled. But because complete recycling without loss is impossible – the law of entropy forbids it as absolutely as perpetual motion – and often also requires vast amounts of energy,[57] the premium should be on longevity, repair, and reuse. Such a conservation economy will still require much research and engineering, but many more menders and fixers. Not so much the Japanese or the German model as the African genius for making things last. The literal anthropological triumph of the *bricoleur* over the engineer.[58] And that may also be a good thing, as long as we remember that bricoleurs generally do not expect to be paid as much as engineers; in fact they normally don't expect to be paid at all, except with thanks. We might still be able to have everyone thankfully employed in a stationary-state economy, and even during an intentionally declining phase in the excessively large-footed parts of the world, but unlike in the previous redistributions of labor-time that have been identified with social progress, we should also accept a consequent, if not necessarily commensurate, *lowering* of our material standard of living. And that may not be a bad thing either, and perhaps, if we can believe the happiness indexes mentioned earlier, more readily acceptable than we think for most people beyond a certain level of sufficiency. But in any case, it is not really negotiable, other than in terms of the eminently political question of how it is to be done. It is a fact of life on earth: compared to a resource-intensive and environmentally destructive economy boosted by fossil fuels, a renewable economy – one living within the limits of the planet and at the rhythm of its naturally perpetuating energy flows – must necessarily have a slower metabolism. For us collectively, whether or not we agree on how to share, it means a lower consumption level, what we ordinarily consider as being poorer.

A closely related question, at least as forbidding, is whether a fully renewable economy – a stationary or steady-state material economy – is compatible with capitalism. Not surprisingly, it is more often raised in those parts of the world where the Marxist tradition is still politically alive, such as Europe and Latin America, and where many people, still at best merely resigned to free-market liberalism, come to environmentalism half hoping that the answer will be negative.[59] But that is no reason not to ask. Common sense as well as basic economic theory dictate that capitalism needs a positive return on investment in order to function, which can only happen if additional value, more wealth, is generated at the end of each investment cycle. Capitalism does not just produce growth as a side benefit; it is consubstantial with it. As we have seen, this is not an issue for the proponents of green or "natural" capitalism, because they believe in massive decoupling, the possibility of continuous economic growth in the absence of any increase in material throughput. But we have also seen that there are reasons to be cautious about their exuberance, and to suspect that the dream of decoupling may be an illusion. The answer as to the viability of the system may lie in a narrow gap between ecological and economic sustainability, and whether both can be maintained at the same time. Reassuringly, even surprisingly, a figure as sober as Herman Daly, whose economic as well as environmental credentials are not to be doubted, seems to think they can. After all, as he reminds us, a steady-state economy (his preferred term for a system with constant throughput, or in ecological equilibrium) is not static and unchanging; it is still active, innovative, and evolving, only within the material limits of lasting sustainability. It is still subject, in keeping with Joseph Schumpeter's famously oxymoronic description of capitalism, to some measure of creative destruction,[60] even if with a lesser emphasis on the latter – improving in quality if not growing in quantity. And no matter how frugal, how conservative we become, there will still be some wear and tear, and scarcity both of resources and capital. In such conditions, qualitative improvements and technical innovations should still be rewarded with higher value and yield some return on investment (Daly, 2008). There may therefore remain enough margin of profit for a capitalist segment of the economy to make a useful contribution and endure indefinitely. But most careful observers recognize that it would not be capitalism as we know it. We should rather think of it, if it is to be sustainable, at best as slow capitalism.

Such a hedging, albeit honest answer, is almost guaranteed to please no one, either intellectually, practically, or ideologically. Those who have long been critical of the social and moral failings of capitalism will argue that the environmental crisis is the final proof of its fundamental non-viability and destructiveness, and that the ecologically stationary state should be founded on a new non-profit and cooperative economic system (some of which is indeed in existence and functioning successfully, though not necessarily ecologically) devoted to the common good. Those who on the contrary are persuaded that capitalism is so integral and indispensable to the modern global economy that its basis cannot be challenged without putting our entire social welfare in grave danger see this fact as a major and inescapable argument for the impossibility of reaching the stationary state – this is indeed the mainstream "realistic" position held by almost all economists and politicians, for whom the pursuit of growth remains the only option – regardless, it should be added, of their increasingly developed and sincere consciousness of environmental limits.

And yet, however unappealing to ideological convictions, even to honorable political commitments, a degree of truthful uncertainty with regards to the ultimate solutions to the challenge of limits is still preferable. Not only because truthfulness should always be preferred in principle – even though it remains a moral luxury political parties, including established Green parties, cannot realistically afford – but also because pragmatically we should really leave all options open, and no person or constituency of good will out of the great shift awaiting us. The transition to a steady state universe, with full acceptance of the end of growth, is so profound and so transformative on so many levels that it calls for all the resources that can be brought to bear: human, technological, and political. If free market solutions can contribute to achieving a fully sustainable life and still provide incentives for private investments, we should welcome them and recognize their merit. When on the other hand market mechanisms manifestly fail to take account of limits, or deliberately abuse them, as is most grievously the case with externalities, but also with so called free-riders (agents who profit from favorable conditions paid for by someone else), then only public authorities can set the necessary regulatory and fiscal framework – which is the way capitalism has always worked best anyway – except with the additional and paramount objective of respecting limits.[61] It is probable that in this new environment the non-profit sector, as well as the governmental and international organizations that have always been designed for the common good (though not traditionally with limits in mind) will have a greater role to play. But there is no reason to think that human institutions and behavior in a steady state economy will be less messy, and even contradictory, than they have been in the past. They will just have to collectively respect the needs of the non-human world.

A little more theoretically, the implications of the stadial model that has been the guiding framework of this reflection should make us wary of expecting whole new worlds appearing under the sun, even less on tables from which the past has been erased.[62] The shift that must be achieved can verily be thought of as a revolution, in the classical sense the term carried until the eighteenth century: a return to an earlier state in a cyclical pattern. We are talking about a modern, technological and informed reinvention of the largely sustainable and very low-growth conditions that prevailed throughout the world before the scientific and industrial revolutions (understood this time in their modern meaning). It could from this perspective be tempting, even inspiring in a grand narrative kind of way to call for the advent of a new stage, the last and truly final stage. For a demanding project that has very little popular attractiveness and could certainly benefit from positive mediatization, such stadial eschatology does have its conceptual appeal. The final struggle for the final stage. But much of what we have learned about the succession of stages suggests that nothing fundamental about the past is really left behind, that stages do not replace but transform one another in their newer forms, and that there never truly is a final stage, just the one we are finally in. It is therefore more accurate, and probably more beneficial too, to think of the achievement of a steady-state economy and the end of growth as the culmination of the stage we have already been in for a good 200 years, the one that brought us industrial technology and fossil fuel energy, and must now provide the resources of the former to gradually displace our reliance on the latter. The last state, and why not, the maturation of an industrial and commercial stage that would finally have found the delicate equipoise of enduring sustainability.

And finally, there remains the ongoing question that has repeatedly surfaced in this reflection, as it would have to in any confrontation with limits: the question of inequality. Not only is the notion of equality so tightly twined with the consideration of limits as to be its inescapable horizon, but it is also – and this quite obviously – so much larger, and more difficult, that it must take us beyond the present focus on ecological limits, as we will have to recognize in the necessary epilogue to follow. And yet, for all its practical and philosophical complexity, and unlike the equally difficult problems evoked in the preceding paragraphs and left in a certain state of indecisiveness, the question of inequality emerges from the test of limits with greater clarity and more moral exigency than ever.

One of the most frequent and compelling arguments in support of continuous growth, and therefore against the possibility of achieving a stationary economy, is the persistence of vast inequalities of wealth between regions and nations, indeed of absolute poverty in much of the world. It is an undeniable fact that within flying and boating distance of unprecedented affluence, billions of people still live in wretched material conditions, often close to actual starvation levels, and barely registering in our collective planetary footprint – though in their desperation and deprivation they too can still be destructive of the few natural resources at their disposal. And billions more, though they have recently escaped from such direst scarcity, are yet very far from a life of modest sufficiency, let alone a fair share of the planetary bounty they largely produce for the rest of us – though they get more than their share of the dregs and refuse of affluence. By any standards of equitability, of mere human decency, they, those of us who represent perhaps two thirds of the world population, need to be able to improve their standard of living, to grow in material wealth, not necessarily, indeed preferably not in the same way as the first world has so far done it, or with the same destructive results, but enough to reach some measure of qualitative fairness in our sharing of the earth. Ecological equality. But that is precisely why those who have already attained more than sufficient material wellbeing, and are almost entirely responsible for the planetary overshoot we are currently experiencing, must imperatively stop growing, and indeed have a moral obligation to reduce their consumption of the world's finite resources. For living within limits changes all the rules. It is what is dismissively called a zero-sum game.

The traditional justification of growth – in orthodox economic terms – is that it eventually increases everyone's wealth. Colloquially, it raises all boats. Over time, even the poorest members of society, though they may remain relatively just as unequal to the wealthiest, will one day reach a level of affluence comparable to that once enjoyed only by their betters. This is the basis for what Fred Hirsch calls "dynamic egalitarianism," according to whose logic the static inequalities prevailing at a given time "provide the incentive as well as the means for general advance" (Hirsch, 1976, 166). The best simile is that of a marching column constantly moving forward, in which ranking represents income distribution. In this version of economic progress, as Michael Young and Peter Wilmott describe it,

> the last rank keeps its distance from the first, and the distance between them does not lessen. But as the column advances, the last rank does eventually reach and pass the point which the first rank had passed some time before.

As a result of this theoretically uniform progression, "the people in the rear cannot, without breaking rank and rushing ahead, reach where the van *is*, but since the whole column is moving forward, they can hope in due course to reach where the van *was*" (Young and Wilmott, 1973, 20). Thus it is clear that if economic growth allows everyone's condition to improve in absolute terms, it will never by itself lessen inequality, but merely make it more retrospectively and prospectively acceptable. In the deservedly well recognized and often quoted formulation of Henry C. Wallish, "growth is a substitute for equality of income. So long as there is growth there is hope and that makes large income differentials tolerable" (Wallish, 1972).

A substitute for equality that is also thereby the alibi of inequality, the main excuse, even justification for the perpetuation and acceptance of social injustice.

As long as there is growth, there is no need for real equity, just for a future semblance of catching up. The achievement of equality among nations and people – a goal we claim to agree upon in principle – can be indefinitely postponed. Which means that we have given up on reaching it, without quite having to admit such a failure of our collective promise. But with the end of growth, the alibi disappears, and the bluff is called. An unequal world without growth is one in which the rich have to tell the poor to their face: yes, it will always be so, you will never have what we have, not now nor at any future time. Or it is, more likely, a world in which the rich will face the risk of being overrun by the multitudes justly outraged at being condemned to perpetual inferiority. By depriving it of its alibi, and thus revealing it as endemic, the absence of growth makes inequality morally intolerable.

But then, as Richard Wilkinson and Kate Pickett point out more positively, the reverse corollary must also be true: "It is not simply that growth is a substitute for equality, it is that greater equality makes growth less necessary. It is a precondition for a steady-state economy" (Wilkinson and Pickett, 2009, 226). If a non-growing world has less tolerance for inequality, a more equal world has less incentive to grow. The less growth, the more equality; the more equality, the less need for growth. Whichever is the chicken, and which the egg, the mutually reinforcing linkage between the level of social inequality and the need for economic growth, with its corollary emphasis on equality in a steadier world, is perhaps the most striking, and certainly one of the most important instances of the human consequences of having to rethink the world in terms of limits.

We have previously evoked some of the other major consequences of living within limits, at least those that come to mind as the most immediate and inescapable. But there are many more, indeed an indefinite range of our choices in all walks of life, that turn out to be affected by, and in turn to have an impact on, the question of limits. For as soon as we begin to accept the reality of finitude and the imperative of abiding by its laws, we come to realize how pervasive its implications are not only in the obvious domains of economics and ecology, but in our cultural, ethical, and indeed in our basic philosophical assumptions. It is not just that so much of our horizon of expectations is premised on growth, and that the withdrawal of quantitative increase as a possibility deprives us of a good part of our usual resorts, whether these were real solutions or largely illusionary. That is merely the negative or privative formulation of what it means to inhabit a limited universe: no, more of the same is no longer an option. But there is the other side, the positive consequence that the only options left open are the qualitative ones, not more but different, and hopefully better (though they could also be worse). And those options remain infinite. Indeed, they may be greatly enhanced in a world without growth. Not only is a steady state society, as we should always remember, not the same thing as a static, unchanging one, but because it can no longer grow in aggregate size, that is to say, in quantity, it is in a sense forced to change in quality, to become qualitatively different. It is therefore not impossible to think that once the necessity of the end of growth is accepted, and its consequences begin to be understood, the achievement of ideals on which we have in practice given up, such as equality among human beings, can again become realistic, and even inescapable, because the alternative will no longer be acceptable.[63]

The end of growth – the only plausible answer to the question of limits – is only a beginning. It raises many other questions, some of which we will now turn to more explicitly, as they cannot in good faith be ignored, though we can by no means address them exhaustively, since such a reflection would be infinite. Some of us will find the prospects thus opened encouraging, and find in the goods they offer further reason to embrace limits. Others may see them as further obstacles and deterrents. Yet even if they mostly do not admit of definite answers, and may in some cases seem extraneous to what should be the main focus for the survival of the world, these wider questions should help us better understand the consequences of living within limits, and the conditions under which it would be best achieved. But one thing should be clear: though the question of ecological limits and our necessary response to it is obviously connected to many other questions and responses, and can be complicated and enriched by those factors, it is not dependent on them for the most fundamental answer, which is both an end and a beginning point. In a finite world, we must stop growing.

Notes

1 As reported by Andrew Blum in *Tubes: A Journey to the Center of the Internet*:

> According to a 2010 Greenpeace report, two percent of the world electricity usage can now be traced to data centers, and that usage is growing at a rate of 12 percent a year. By today's standards, a very large data center might be a 50,000-square foot building demanding fifty megawatts of power, which is about how much it takes to light a small city.
>
> (Blum, 2012, 230)

2 As acknowledged in Chapter 1, Brian Czech has labeled the same illusion "The Fallacy of Self-Sufficient Services" (Czech, 2009), and offers a powerful critique using the biological model of trophic levels. His work will be recognized more fully after our own exposition in terms of stadial and sectoral progression.

3 The theory of sectoral growth is primarily associated with Colin Clark (1909–1989), with contributions by Alan Fisher and Jean Fourastié. It is also related to the influential theorizations of Simon Kuznets, whom we will encounter later. Sectoral economic analysis is sometimes labeled structuralist, which could be confusing for those more familiar with the humanities, where structuralism, linguistic or otherwise, means something altogether different. See W. W. Rostow: *Theorists of Economic Growth* (Rostow, 1990, Chapters 13 and 16).

4 As we noted in the previous chapter, in 2017 the percentage of people employed in agriculture was 3 percent in high-income countries and in the Euro area, 2 percent in the US, Canada and the Netherlands, and 1 percent in the UK and Germany (World Bank, 2017). In terms of contribution to GDP, the percentage is even lower (the World Bank data in this case combines the value added of agriculture, forestry and fishing): 1.5 percent for the Euro area, 1 percent for North America, 0.6 percent for Germany, and 0.5 percent for the UK (World Bank, 2018).

5 One of the difficulties with making any generalizations about the service sector is that it is so vast and disparate. As an example, this is the list of service activities reproduced by Peter A. Victor from official Canadian statistics:

> Wholesale and retail trade; transportation and warehousing; information and cultural industries; finance, insurance and real estate services; professional, scientific and technological services; administrative and support, waste management and remediation services; education; health care and social assistance; arts, entertainment and recreation; accommodation and food services; other services; and public administration (Statistics Canada, Cansim Table 379-0017).
>
> (Victor, 2008, 105)

6 In *Adam Smith's Discourse* (Brown, 1994, 177–178).
7 Although I am pleased to find that Aldo Leopold, who was a biologist as well as a pioneering environmentalist, does recall the same principles in *A Sand County Almanac*, though not about economic sectors (Leopold, 1949, 175).
8 See for example Roland Oliver, *The African Experience* (Oliver, 1991, 32). Jared Diamond has also insisted on this slow process in *Guns, Germs and Steel* (Diamond, 1997).
9 Daniel Bell writes:

> a post-industrial society does not "displace" an industrial society, or even an agrarian society. Food is still the foundation of all society, but what the introduction of industry meant was that one could reduce the number of persons engaged in agriculture and increase the yields because of chemical fertilizers.
>
> (Bell, 1996, 198)

10 Although W. Arthur Lewis in his *Theory of Economic Growth* makes the striking remark:

> The fact is that the great inventions of the eighteenth and the nineteenth centuries … were all invented by practical people who knew no science, or very little. It is only in the twentieth century that a scientific education has become essential for the would-be inventor, or that the discoveries of science have become the major source of further technological progress.
>
> (Lewis, 1955, 79)

11 Regarding the industrialization of agriculture, Zolotas notes that

> whereas formerly man used the inflow of the photosynthesis of chlorophyll (i.e. terrestrial flora) as the moving force of oxen, today, in order to move tractors, he is using the depletable resources of terrestrial energy (oil), i.e. energy of high entropy.
>
> (Zolotas, 1981, 121)

Worse, as David Montgomery notes,

> The USDA estimates that half the fertilizer used each year in the United States simply replaces soil nutrients lost by topsoil erosion. This puts us in the odd position of consuming fossil fuels – geologically one of the rarest and most useful resources ever discovered – to provide a substitute for dirt – the cheapest and most widely available agricultural input imaginable.
>
> (Montgomery, 2007, 200)

12 Overall, Atsuchi Tsuchida and Tadeshi Murota write,

> the propagation of information processing machines may increase the consumption of available matter and/or energy in the economy, instead of decreasing it. This may result in the intensification of underground materials' pollution, from which the said information society is hoped to be free.
>
> (Pilet and Murota, 1987, 27)

13 Paul Burkett notes:

> As Huws observes, the current production of ever greater amounts of information and entertainment (the main locus of today's service economy) is dependent on

> the growing matter-energy throughput associated with computers, scanners, printers, mobile phones, media players, disks, and so forth, which are subject to more and more rapid rates of obsolescence – not to mention the "many components and accessories involved in their manufacture."
>
> (Burkett, 2006, 159)

The quotation is from Ursula Huws: "Material World: The Myth of the Weightless Economy" (Panitch and Leys, 1999, 49).

14 See note 4 (this chapter).

15 According to Karen Higgins: "Agriculture drinks up an estimated two-thirds of the world's water [...] It takes 1,000 tons of water to grow a ton of wheat, 65 gallons per pound of potatoes, and 650 gallons per pound of rice" (Higgins, 2015, 84).

16 The United States Department of Agriculture (October 14, 2016) estimates it at 10 percent of total US greenhouse gas emissions, dominated by nitrous oxide and methane. This compared to its approximately 1 percent share of national GDP (USDA, 2016).

17 As Xenophon Zolotas puts it: "... the function of industrial development is to convert rapidly the valuable and finite stock of free terrestrial energy (low entropy) into bound energy (high entropy)." He further dismisses a common illusion about a way to escape from the problem: "Recycling cannot in the long run counterbalance the second law of thermodynamics, since use will again be made of free energy which will turn into bound energy." Of course, as he also acknowledges, recycling can delay the entropic process and postpone "the depletion of the finite stock of energy sources," but it cannot do so forever (Zolotas, 1981, 56–57).

18 Summarizing a study published by the National Academy of Sciences on the emissions of industrialized countries that have signed the Kyoto Protocol: "Growth in Emission Transfers via International Trade from 1990 to 2008" (Peters *et al.*, 2011), Naomi Klein writes:

> It found that while their emissions had stopped growing, that was partly because international trade had allowed those countries to move their dirty productions overseas. The researchers concluded that the rise in emissions from goods produced in developing countries but consumed in industrialized ones was six times greater than the emissions savings for industrialized countries.
>
> (Klein, 2014, 80)

Also see Richard Heinberg, *The End of Growth* (Heinberg, 2011, 166–168), and Adam Minter: *Junkyard Planet* (Minter, 2013).

19 Brian Czech's favorite example of a truly environmentally offensive "service" activity, included in what we might ordinarily think as the relatively benign "entertainment sector," is NASCAR (National Association for Stock Car Auto Racing). An extreme example, perhaps, but then as he points out: "It's fair enough when we consider that NASCAR has been the fastest-growing spectator sport in North America over the past several decades" (Czech, 2013, 38).

20 As summarized by Burkett: "Kuznets posited that starting at low per capita income levels, income inequality tends to rise as the economy grows, but then eventually reaches a turning point beyond which inequality falls as per capita income rises" (Burkett, 2006, 210). See Simon Kuznets, "Economic Growth and Income Inequality" (Kuznets, 1955).

21 This is clearly articulated by Benjamin M. Friedman:

> The relationship between economic development and environmental degradation may therefore follow a pattern analogous to what Simon Kuznets years ago posited as the relationship between development and inequality: In the initial stages of development, the transition from traditional agriculture to manufacturing increases pollution and greenhouse gas emissions. But in time further shifts

> in the composition of economic activity (mostly toward a larger service sector), the evolution of new technology, and both private and public choices that higher income make possible, all lead instead to an improvement.
>
> (Friedman, 2005, 382)

22 We should note, however, that transportation only accounts for 14 percent of global GHG emissions. While making cars more fuel efficient helps reduce carbon dioxide emissions, it may not be the largest factor in countries becoming less carbon-intensive. The sector that contributes the most to global greenhouse gas emissions is "electricity and heat production (25 percent)," followed closely by agriculture, forestry and other land use (24 percent). Reducing the carbon intensity of these activities (such as generating electricity from renewable sources and amending land use practices) has the most potential for decreasing global greenhouse gas emissions in the future and effecting the CO_2 levels in the present (Environmental Protection Agency, 2017).

23 Electricity use seems to confirm this positive trend. Following a World Bank chart that tracks per capita consumption by country, and taking for comparison changes between 1971 to 2014, we see that while developing countries show uninterrupted steep or exponential growth, such as China from 152 kWh to 3,927 kWh, and India from 98 to 806, highly developed countries, after a period of steady growth, reach a plateau during the years 2000–2010, and then slowly decline. Thus the OECD countries from 3,957 kWh in 1971 to a maximum of 8,391 in 2008 and down to 7,994 in 2014, and the US, from 7,517 kWh to 13,671 in 2000 and 12,984 in 2014. If the decline in OECD countries follows the recession starting in 2007, this is not the case in the US (World Bank, 2014).

24 As Brian Czech writes: "There is no evidence for a macroeconomic environmental Kuznets curve. Many sector- or industry-specific environmental problems have been created and exacerbated as a function of economic growth; few have exhibited a Kuznets curve" (Czech, 2013, 201). Czech refers to David I. Stern: "The Rise and Fall of the Environmental Kuznets Curve" (Stern, 2004).

25 Brian Czech's first book was *Shoveling Fuel for a Runaway Train* (Czech, 2000), already cited. "The Fallacy of Self-Sufficient Services" was the title of a short piece published in 2009 in *Frontiers in Ecology and The Environment* (Czech, 2009).

26 It could be pointed out that this bears a strong resemblance to what Aldo Leopold, another biologist, had much earlier called "the land pyramid" (Leopold, 1949, 215).

27 Brian Czech does not indicate at this point whether some humans should be included among the "super-carnivores," but there are indications elsewhere in his work, for example Chapter 7 of *Shoveling Fuel*, "Relations with the Liquidating Class" (Czech, 2000).

28

> The economic process (production followed by consumption) is entropic. Raw materials from nature are equal in quantity to the waste materials ultimately returned to nature [First law of thermodynamics]. But there is a qualitative difference between the equal quantity of raw and waste material [Second law]. Entropy is the physical measure of that qualitative difference.
>
> (Daly and Cobb, 1989, 195)

29 The full title of the chapter, in Czech's often earthy language, is "Don't Sell the Farm: The Trophic Theory of Money." Why we shouldn't sell the farm should now be obvious: without the farm, no money (Czech, 2013).

30 Czech argues,

> In the human economy, the service sectors [...] defy placement in a particular trophic level. Truck drivers, bankers, waitresses, janitors, gravediggers – none produce or consume in a systematic fashion that proceeds upward from one trophic level to the next. The truck driver may deliver a load of cotton from farm

to factory one day, and a load of fence from factory to farm the next. The banker may lend to the farmer or to the industrialist.

(Ibid., 177)

31

> [J]ust as the capacity of the economy of nature is based on the amount of primary production, the capacity of the human economy is based on the amount of natural capital. Within the economy, the production of the manufacturing trophic level is dependent on the production of the primary producers – the farmers, miners, loggers and such. The service providers depend on the whole system.
>
> (Ibid., 182)

32 The first edition of *The Limits to Growth* appeared in 1972 (Meadows *et al.*, 1972); the first anthology of essays by Herman E. Daly: *Toward a Steady State Economy*, in 1973 (Daly, 1973) and his first book still in print: *Steady State Economics*, in 1977 (Daly, 1990). In *Prosperity without Growth*, Tim Jackson's references include *The Steady State Economy*, London, W. H. Freeman & Co., Ltd as published in 1972, but I have not found any confirmation of its existence (Jackson, 2009, 243).

33 "Today humanity uses the equivalent of 1.5 planets to provide the resources we use and absorb our waste. This means it now takes the Earth one year and six months to regenerate what we use in a year" (footprintnetwork.org, September 1, 2011). Citing the WWF and the Global Footprint Network *Living Planet Report 2008*, Paul Gilding adds: "Based on their own data sets and projections, the Global Footprint Network has warned that we are on course to be running at 200 percent of capacity by the early 2030s – two planets worth" (Gilding, 2011, 51–52).

34 The average American's ecological footprint is approximately 9.7 hectares per person per year, while the global capacity is 1.8 hectares per person. In other words, if everyone in the world lived like an average American, we would need almost five and a half planets (9.7/1.8 = 5.38) to support us in the long run.

(Czech, 2013, 269)

35 Bill McKibben writes:

> Given current rates of growth in the Chinese economy, the 1.3 billion residents of that nation alone will, by 2031, be almost as rich as we are. If they eat meat, milk, and eggs in the same quantities as we do [...] they'll consume 1,352 million tons of grain, or two thirds of the world's entire grain harvest. They'll use 99 million barrels of oil a day, 20 million more barrels than the entire world consumes at present [...] They'd use more steel than all the West combined. Paper? At the American rate, they'd consume 303 million tons, roughly double the current world production. Cars? They'd have 1.1 billion on the road, half again as many as the current world total.
>
> (McKibben, 2007, 18–19)

36 In 2008, Martin Rees was writing:

> To ensure a better than even chance of avoiding a potentially dangerous "tipping point," global CO_2 emissions must, by 2050, be brought down to half the 1990 level. This is the target espoused by the G8 [the very conservative group of most industrialized nations that at the time still included Russia]. It corresponds to two tons of CO_2 per year from each person on the planet. For comparison, current CO_2 emissions for the US are about 20 tons per person, the average European figure is about 10, and the Chinese level is already 4 [tons].
>
> (Rees, 2008, 41)

In a 2004 "Carbon Mitigating Initiative," Robert Socolow and Stephen Pacala proposed a list of 15 measures, 8 of which would have to be implemented to bring CO_2 levels down to 280 parts per million. It is too long to be reproduced here, but quite sobering, even if we recognize that the potential for solar and wind power has since then improved greatly (Friedman, 2008, 212–213). See also the equally demanding "One Degree War Plan" proposed in 2009 by Jorgen Randers and Paul Gilding (Randers and Gilding, 2010).

37 Stephen Schneider writes:

> The dangers associated with climatic change are merely a part of the entire world *problématique*, and to view them in isolation from the rest of the world predicament would be to repeat the mistakes of many narrowly specialized observers who have examined the prospects for the future *only through the tunnel of their experience*.
>
> (Schneider, 1976, 246)

38 "Biologist E. O Wilson estimates that the Earth is losing about one species every 20 minutes, or almost 30,000 species a year, 1,000 times faster than the normal over the history of the earth" (Edwards, 2010, 92). According for the Center for Biological Diversity, "scientists estimate we're now losing species at 1,000 to 10,000 times the background rate, with literally dozens going extinct every day [...] as many as 30 to 50 percent of all species possibly heading toward extinction by mid-century" (Center for Biological Diversity, n.d.).

39 In "Understanding the Butterfly Effect," James L. Vernon informs us that the concept was introduced by MIT meteorology professor Edward Lorenz, who queried in 1972:

> Does the flap of a butterfly's wings in Brazil set off a tornado in Texas? [...] The purpose of his provocative question, he said, was to illustrate the idea that some complex dynamical systems exhibit unpredictable behaviors such that small variances in the initial conditions could have profound and widely divergent effects on the system's outcomes. Because of the sensitivity of these systems, outcomes are unpredictable.
>
> (Vernon, 2017, 130)

40 "The Study of Earth as an Integrated System," a short article available on the NASA Science website, provides a clear summary with examples of major climate feedback loops, defined as "processes that can either amplify or diminish the effects of climate forcings" (clouds, precipitation, greening of the forests, ice albedo). It also summarizes the major forcings, defined as "the initial drivers of the climate" (solar irradiance, greenhouse gas emissions, aerosols, dust, smoke and soot). The related concept of tipping points is defined as "when Earth's climate abruptly moves between relatively stable states," and gives examples of how these tipping points are reached (changes in ocean circulation, ice loss, and rapid release of methane) (NASA, n.d.).

41 The title of a 1980 book by William R. Catton: *Overshoot: The Ecological Basis of Revolutionary Change* (Catton, 1980).

42 At least to a French ear. Degrowth is the literal translation of *décroissance*, the opposite of *croissance*. But unlike in English, where even the word growth itself isn't particularly lovely, and its negative rather unnatural, *croissance* and *décroissance* in French are as familiar and reassuring as the waxing and waning of the moon, which *croît* and *décroît* every month, forming in the process a shape known by its French name by pastry lovers even in English.

43 Author with Amory Lovins and Hunter L. Lovins of *Natural Capitalism: Creating the Next Industrial Revolution* (Hawken *et al.*, 1999), and on his own or as editor, of many other inspiring books.

44 Aside from renewable energy and changes in forestry and agriculture, the areas of greatest environmental impact, there are many other technological investments that

could help lessen our footprint, and therefore increase decoupling. In the recently published comparative study of such solutions, *Drawdown*, edited by Paul Hawken, specifically focused on global warming, the technological change identified as number one in terms of potential reduction is, astonishingly until we think about it, "refrigerant management" (Hawken, 2017).

45 There is a considerable weight of argument on the negative side of the decoupling debate. According to Clive Hamilton,

> even to keep the impact of economic activity of Earth at its current destructive level, [over the next 50 years] technological advance will have to reduce resource use per unit of output to but a quarter of the current levels.
>
> (Hamilton, 2004, 176)

Citing a Michigan State University study (Dietz *et al.*, 2007), Brian Czech writes: "If, as frequently suggested, technological progress can redress environmental problems [...] the requisite technological improvement needs to exceed two percent per year" (Czech, 2013, 208). This is a level of constant productivity gains that has not been achieved since the third quarter of the twentieth century. See also the entire fifth chapter of Tim Jackson's *Prosperity without Growth*, "The Myth of Decoupling," and his conclusion in a later chapter: "In a world of nine billion people all aspiring to western lifestyles, the carbon intensity of every dollar of output must be at least 130 times lower in 2050 than it is today" (Jackson, 2009, 187).

46 An obvious counter to this trend is E. F. Schumacher's *Small is Beautiful*, first published in 1973, a date in the modern history of limits (one year after the first *Limits to Growth* and the same as Herman E. Daly's *Toward a Steady State Economy*) we can now look back to with a degree of nostalgia.

47 The World Bank charts the decline of fertility rates (average number of births per woman) between 1960 and 2016. For the whole world it fell from almost 5 down to 2.439. As well as data for each country, the site provides a useful summary by significant groupings of countries. Thus between the same dates fertility in the EU declined from 2.6 to 1.6, in heavily indebted poor countries from 6.7 to 4.9), in "Least developed countries – UN classification" from 6.7 to 4.1, in OECD member countries from (3.2 to 1.7), in high income countries from 3.0 to 1.7, in middle income countries from 5.6 to 2.3), and in low income countries from 6.6 to 4.6 (World Bank, n.d.).

48 The rebound effect normally describes the phenomenon whereby gains in efficiency result in an increase in use, so that the potential reductions in energy or resource consumption are eventually cancelled, for example when improvements in fuel efficiency in automotive technology encourage people to buy bigger or more cars instead of saving on petroleum. Its first formulation is often attributed to William Stanley Jevons, who wrote in *The Coal Question* (1865): "It is wholly a confusion of ideas to suppose that the economical use of fuel is equivalent to a diminished consumption. The very contrary is the truth" (Dietz and O'Neil, 2013, 39).

49

> According to calculations worked out by Paul and Anne Ehrlich, a child born [in 1990] in the United States will yield a destructive force 280 times that of its counterpart in Nepal, Haiti, or much of Africa, 140 times that of a Bangladeshi, 35 times that of an Indian Child, 13 times that of a Brazilian.
>
> (Ehrlich and Ehrlich, 1990, 134)

50 Also known as "Downshifting." Tim Jackson summarizes a number of surveys with striking results:

> The downshifting movement now has a surprising allegiance across a number of developed economies. A recent survey on downshifting in Australia found that 23 percent of respondents had engaged in some form of downshifting in the five

> years prior to the study. A staggering 83 percent felt that Australians are too materialistic. An earlier study in the US found that 28 percent had taken some steps to simplify and 62 percent expressed a willingness to do so. Very similar results have been found in Europe.
>
> (Jackson, 2009, 150–151)

See also Jim Merkel, *Radical Simplicity* (Merkel, 2003).

51 Richard Wilkinson and Kate Pickett cite a report by the Hardwood Group Institute for Public Innovation: *Yearning for Balance* (1995), concluding that a "large majority of Americans wanted society to 'move away from greed and excess,' but felt 'these priorities were not shared by most of their fellow Americans'" (Wilkinson and Pickett, 2009, 4).

52 Avner Offer writes:

> The positive cross-sectional correlation of income and [Subjective Well Being] *within* countries has long suggested a link [...] to the "relative income hypothesis," which states that what counts is not absolute income, but relativities. From this point of view, even a large rise in income will leave no impart on well-being, if distribution is unchanged.
>
> (Offer, 2006, 31–32)

Similarly, citing Richard Layard's *Happiness: Lessons for a New Science* (Layard, 2005), Peter A. Victor confirms that "... real GDP per capita more than doubled in the USA in the half century following the end of the Second World War but the percentage of Americans who described themselves as 'very happy' hardly changed." And he comments astutely: "Americans have been more successful decoupling GDP from happiness than in decoupling it from material and energy" (Victor, 2008, 125).

53 This would be a place to develop what could be called, following the other side of Adam Smith, an economy of sympathy. It is also an opportunity to acknowledge the argument of Arran E. Gare, who sees in the human need for recognition a possible counterweight to the competitive social dynamic emphasized by hierarchy theorists. Going back to the Greek *polis* and following the lineage from Herder and Hegel to Habermas, Gare insists that the "struggle for recognition" exercises a beneficial restraint on the competitive struggle. "Could the struggle for recognition and the quest for justice which has issued from it [...] provide the constraints in human interactions with nature and on human relationships which could divert humans from the present destructive path of development?" Responding to the revolutionary call by Ilya Progogine "to create a new alliance between science and the humanities and 'humanistic' social sciences," Gare contends that

> [t]he development of human (or social) ecology is central to this revolution. With [it] it becomes possible to understand the ethical development of humanity as a kind of constraint which could limit the present environmentally destructive trajectory of civilization...
>
> (Gare, 2000, 287–288)

54 As usual, Tim Jackson is particularly lucid in this respect, placing his hopes for a steady state society in the development of the personal and social services sector, where labor productivity gains are lower (Jackson, 2009, 63; 132), and praises Peter A. Victor's policy recommendations along these lines for Canada in *Managing without Growth*, Chapter 10. Advocates of *Décroissance* in Europe have long made reduction of labor time central to their vision, see among many others Serge Latouche: *Le pari de la décroissance* (Latouche, 2006).

55 Here is the dire verdict of Joseph Tainter:

> Given the close link between economic and military power, unilateral economic deceleration would be equivalent to, and as foolhardy as, unilateral disarmament.

> We simply do not have the option to return to a lower economic level, at least not at a rational level.
>
> (Tainter, 1988, 214)

56 In 1976, Daniel Bell predicted:

> By the end of the [twentieth] century, a single space-time framework will finally be in place, engirdling the entire world. We will have reached the great *Oekumene* – the single household economy – which the Greeks had envisioned as the boundary of the civilized world.

And he asked:

> How much can be managed from a single center? How large can a political or economic enterprise be, without becoming a behemoth that falls of its own weight? How many nations can be effective participants in a world assembly? The major question before the international society is the construction of new forms proportionate to the scale on which we live.
>
> (Bell, 1996, 212)

Starting from similar reference to the Greek world and the Hegelian hope for "global civilization with global institutions able to constrain people," Aran E. Gare reaches the opposite conclusion: the necessity to "decentralize political power and insulate as much as possible the economies of different parts of the world [...] An international order working to minimize economic interactions between blocks of nations" (Gare, 2000, 289).

57 Paul Burkett provides good summary of the most authoritative opinion on the question:

> While taking an agnostic stance on the *physical impossibility* of complete matter-recycling, [Herman E.] Daly is also "prepared to believe in common-sense evidence that for all practical purposes complete recycling is impossible." Like Georgescu-Roegen, he points to "the physical fact that enormous amounts of energy, as well as other materials, are required to recycle highly dispersed matter." It "thus remains clear that complete materials recycling would require ruinous amounts of energy and time."
>
> (Burkett, 2006, 155–156)

Jon Elster, also referring to Georgescu-Roegen's *The Entropy Law and the Economic Process*, is if anything even more conclusive:

> profit, interest and economic growth are possible only because man can tap external sources of raw material and energy. It follows from the second law of thermodynamics that an economy based exclusively on recycling would have a negative growth rate.
>
> (Elster, 1985, 141)

58 The distinction between the engineer and the *bricoleur* comes from Claude Lévi-Strauss, who in *The Savage Mind* (1962) describes the latter as the tinkerer who uses anything that happens to be at hand, without previous design, to fix whatever problem presents itself, as opposed to the engineer's methodical and scientific approach (Lévi-Strauss, 1966).

59 This is certainly the case for the militant French organization Attac (Association pour la taxation des transactions financières et pour l'action citoyenne), and for many who call themselves *altermondialistes*. But even if less radically, it is also the position shared by Naomi Klein in *This Changes Everything*, as her subtitle, *Capitalism vs. The Climate*, make fairly explicit (Klein, 2014).

60 The concept of creative destruction was developed by Schumpeter in *Capitalism, Socialism and Democracy* (Schumpeter, 1942), and remains his most identifiable legacy, aside from his formidable *History of Economic Analysis* (Schumpeter, 1954).

61 Witness the vigorous debate among committed environmentalists between the advocates of cap and trade solutions and the partisans of taxation to address emissions of all kinds, including the prominent case of global warming gases.
62 The literal meaning of *tabula rasa*.
63 I am happy, for once, to quote from one of my other compatriots, Jacques Généreux, in my approximate translation:

> There remains a long way to go before we admit that the necessary growth of material conditions in poor countries necessitates, in order to be sustainable on a planetary scale, their *décroissance* in rich countries. A long way until we draw all the conclusions from this elementary observation: a person living in comfortable material conditions can continually expand her well-being through culture, wonder, sharing, love, conviviality, and the quest for meaning, all things that consume nothing but an infinitely reproducible human energy.
>
> (Généreux, 2002, 68–69)

References

Bell, Daniel (1996) *The Cultural Contradictions of Capitalism*. New York: Basic Books.

Blum, Andrew (2012) *Tubes: A Journey to the Center of the Internet*. New York: Harper Collins.

Brown, Vivienne (1994) *Adam Smith's Discourse: Canonicity, Commerce and Conscience*. London: Routledge.

Burkett, Paul (2006) *Marxism and Ecological Economics: Toward a Red and Green Political Economy*. Leiden: Brill.

Catton, William R. (1980) *Overshoot: The Ecological Basis of Revolutionary Change*. Urbana and Chicago: University of Illinois Press.

Center for Biological Diversity (n.d.) "The Extinction Crisis" [online]. Available at www.biologicaldiversity.org/programs/biodiversity/elements_of_biodiversity/extinction_crisis/.

Czech, Brian (2000) *Shoveling Fuel for a Runaway Train: Errant Economists, Shameful Spenders, and a Plan to Stop Them All*. Berkeley, Los Angeles and London: University of California Press.

Czech, Brian (2009) "The Self-Sufficient Services Fallacy," *Frontiers in Ecology and the Environment*, vol. 7, no. 5, 240–241.

Czech, Brian (2013) *Supply Shock: Economic Growth at the Crossroads and the Steady State Solution*. Gabriola Island, DC, Canada: New Society Publishers.

Daly, Herman E. (ed.) (1973) *Toward a Steady State Economy*. San Francisco: W. H. Freeman.

Daly, Herman E. (1990) *Steady State Economics*. Washington, DC: Island Press.

Daly, Herman E. (2008) *A Steady-State Economy*. Sustainable Development Commission [online]. Available at www.sd-commission.org.uk/data/files/publications/Herman_Daly_thinkpiece.pdf.

Daly, Herman E. and John B. Cobb, Jr. (1989) *For the Common Good: Redirecting the Economy toward Community, the Environment, and a Sustainable Future*. Boston: Beacon Press.

Diamond, Jared (1997) *Guns, Germs, and Steel: The Fates of Human Societies*. New York and London: W. W. Norton.

Diamond, Jared (2011) *Collapse: How Societies Choose to Fail or Succeed*. New York: Penguin.

Dietz, Rob and Dan O'Neill (2013) *Enough Is Enough: Building a Sustainable Economy in a World of Finite Resources*. San Francisco: Barrett-Koehler Publishers.

Dietz, Thomas, Eugene A. Rosa and Richard York (2007) "Driving the Human Ecological Footprint," *Frontiers in Ecology and the Environment*, vol. 5, no. 1, 13–18.

Edwards, Andrés R. (2010) *Thriving beyond Sustainability: Pathways to a Resilient Society*. Gabriola Island, BC: New Society Publishers.

Ehrlich, Paul R. and Anne H. Ehrlich (1990) *The Population Explosion*. New York: Touchtone Books, Simon & Schuster.

Elster, Jon (1985) *Making Sense of Marx*. Cambridge, UK: Cambridge University Press.

Environmental Protection Agency (2017) "Global Greenhouse Gas Emissions Data," [online]. Available at www.epa.gov/ghgemissions/global-greenhouse-gas-emissions-data.

Ferkiss, Victor (1993) *Nature, Technology and Society*. New York: New York University Press.

Friedman, Benjamin M. (2005) *The Moral Consequences of Economic Growth*. New York: Vintage Books.

Friedman, Thomas L. (2008) *Hot, Flat, and Crowded: Why We Need a Green Revolution – and How It Can Renew America*. New York: Farrar, Strauss & Giroux.

Gadrey, Jean (2011) *Adieu à la croissance: Bien vivre dans un monde solidaire*. Paris: Les Petits Matins.

Gadrey, Jean and Florence Jany-Catrice (2006) *The New Indicators of Well-Being and Development*. Houndmills, Basingstoke: Palgrave Macmillan.

Gare, Arran E. (2000) "Is it Possible to Create an Ecologically Sustainable World Order: The Implications of Hierarchy Theory for Human Ecology," *International Journal of Sustainable Development and World Ecology*, vol. 7, no. 4, 277–290.

Généreux, Jacques (2001–2002) *Les vraies lois de l'économie*. Paris: Seuil.

Georgescu-Roegen, Nicholas (1971) *The Entropy Law and the Economic Process*. Cambridge, MA and London: Harvard University Press.

Gilding, Paul (2011) *The Great Disruption: Why the Climate Crisis Will Bring on the End of Shopping and the Birth of a New World*. New York: Bloomsbury Press.

Hamilton, Clive (2004) *Growth Fetish*. London: Pluto Press.

Hawken, Paul (ed.) (2017) *Drawdown: The Most Comprehensive Plan Ever Proposed to Reverse Global Warming*. New York: Penguin Books.

Hawken, Paul, Amory Lovins and L. Hunter Lovins (1999) *Natural Capitalism: Creating the Next Industrial Revolution*. Boston, New York and London: Little, Brown and Company.

Hegel, G. W. F. (1807) *The Phenomenology of Spirit*. Oxford, New York, Toronto and Melbourne: Oxford University Press (this edition 1977).

Heinberg, Richard (2011) *The End of Growth: Adapting to Our New Economic Reality*. Gabriola Island, BC, Canada: New Society Publishers.

Higgins, Karen L. (2015) *Economic Growth and Sustainability: Systems Thinking for a Complex World*. Academic Press, Elsevier.

Hirsch, Fred (1976) *Social Limits to Growth*. Cambridge, MA: Harvard University Press.

Howe, Joshua P. (2014) *Behind the Curve: Science and the Politics of Global Warming*. Seattle and London: University of Washington Press.

Jackson, Tim (2009) *Prosperity without Growth*. London and New York: Earthscan-Routledge.

Jevons, William Stanley (1905) *The Coal Question*. New York: Augustus Kelley.

Keynes, John Maynard (2009) *Essays in Persuasion*. New York: Classic House Books.

Klein, Naomi (2014) *This Changes Everything: Capitalism vs. the Climate*. New York: Simon & Schuster.

Kuznets, Simon (1955) "Economic Growth and Income Inequality," *American Economic Review*, vol. 45, no. 1, 1–28.

Layard, Richard (2005) *Happiness: Lessons from a New Science*. London: Allen Lane.

Latouche, Serge (2006) *Le pari de la décroissance*. Paris: Fayard.

Leopold, Aldo (1949) *A Sand County Almanac, and Sketches Here and There*. New York and Oxford: Oxford University Press.

Lévi-Strauss, Claude (1966) *The Savage Mind*. Chicago: The University of Chicago Press.

Lewis, W. Arthur (1955) *The Theory of Economic Growth*. London and New York: Routledge.

McKibben, Bill (2007) *Deep Economy: The Wealth of Communities and the Durable Future*. New York: Holt.

McKibben, Bill (2010) *Eaarth: Making a Life on a Tough New Planet*. New York: Holt.

Meadows, Donella H., Dennis L. Meadows, Jørgen Randers and William W. Behrens (1972) *The Limits to Growth: A Report for the Club of Rome's Project on the Predicament of Mankind*. New York: Universe Books.

Merkel, Jim (2003) *Radical Simplicity: Small Footprints on a Finite Earth*. Gabriola Island, BC: New Society Publishers.

Minter, Adam (2013) *Junkyard Planet: Travels in the Billion-Dollar Trash Trade*. New York, London, New Delhi and Sydney: Bloomsbury Press.

Montgomery, David R. (2007) *Dirt: The Erosion of Civilizations*. Berkeley, CA, Los Angeles and London: University of California Press.

NASA (n.d.) "The Study of Earth as an Integrated System" [online]. Available at https://climate.nasa.gov/nasa_science/science/.

Offer, Avner (2006) *The Challenge of Influence: Self-Control and Well-Being in the United States and Britain since 1950*. Oxford and New York: Oxford University Press.

Oliver, Roland (1991) *The African Experience: Major Themes in African History from Earliest Times to the Present*. New York: Harper Collins.

Panitch, Leo and Colin Leys (eds.) (1999) *Socialist Register 1999: Global Capitalism Versus Democracy*. New York: Monthly Review Press.

Peters, Glen P., Jan C. Minx, Christopher L. Weber and Ottmar Edenhofer (2011) "Growth in Emission Transfers via International Trade from 1990 to 2008," *Proceedings of the National Academy of Sciences*, vol. 108, no. 21, 8903–8908.

Pilet G. and T. Murota (eds.) (1987) *Environmental Economics: The Analysis of a Major Interface*. Geneva: R. Leimgruber.

Prigogine, Ilya and Isabelle Stengers (1984) *Order Out of Chaos*. Toronto: Bentham.

Randers, Jorgen and Paul Gilding (2010) "The One-Degree War Plan." *Journal of Global Responsibility*, vol. 1, no. 1, 170-188.

Rees, Martin (2008) "Science: The Coming Century," *The New York Review of Books*, November 20.

Rostow, W. W. (1990) *Theorists of Economic Growth from David Hume to the Present: With a Perspective on the Next Century*. New York and Oxford: Oxford University Press.

Schneider, Stephen with Lynn H. Mesirow (1976) *The Genesis Strategy: Climate and Global Survival*. New York: Springer Science and Business Media.

Schumacher, E. F. (2010) *Small Is Beautiful: Economics as if People Mattered*. New York: Harper Perennial.

Schumpeter, Joseph A. (1942) *Capitalism, Socialism and Democracy*. New York: Harper and Rowe.

Schumpeter, Joseph A. (1954) *History of Economic Analysis*. New York: Oxford University Press.

Schweiger, Larry J. (2009) *Last Chance: Preserving Life on Earth*. Golden, CO: Fulcrum.

Stern, David I. (2004) "The Rise and Fall of the Environmental Kuznets Curve," *World Development*, vol. 32, no. 8, 1419–1439.

Tainter, Joseph (1988) *The Collapse of Complex Societies*. Cambridge, UK: Cambridge University Press.

Tobias, Michael (1994) *World War III: Population and the Biosphere at the End of the Millennium*. Santa Fe, New Mexico: Bear and Company.

USDA (2016) "Agriculture and Climate Change" [online]. Available at www.ers.usda.gov/topics/natural-resources-environment/climate-change/agriculture-and-climate-change/.

Vernon, Jamie L. (2017) "Understanding the Butterfly Effect," *American Scientist*, vol. 105, no. 3, 130 [online]. Available from (www.Americanscientist.org/article/understanding-the-butterfly-effect).

Victor, Peter A. (2008) *Managing without Growth: Slower by Design, Not Disaster*. Cheltenham, UK and Northampton, MA: Edward Elgar.

Wallish, Henry C. (1972) "Zero Growth," *Newsweek*, January 24.

Wilkinson, Richard and Kate Pickett (2009) *The Spirit Level: Why Greater Equality Makes Societies Stronger*. New York: Bloomsbury Press.

World Bank (2014) "Electric Power Consumption Per Capita" [online]. Data © International Energy Agency www.iea.org. Available at https://data.worldbank.org/indicator/EG.USE.ELEC.KH.PC?view=chart.

World Bank (2017) "Employment in Agriculture Statistics" [online]. Available at https://data.worldbank.org/indicator/SL.AGR.EMPL.ZS.

World Bank (2018) "Agriculture, Forestry, and Fishing, Value Added" [online]. Available at https://data.worldbank.org/indicator/NV.AGR.TOTL.ZS.

World Bank (n.d.) "Fertility Rate, Total (Births Per Woman)" [online]. Available at https://data.worldbank.org/indicator/SP.DYN.TFRT.IN?view=chart.

Young, Michael and Peter Wilmott (1973) *The Symmetrical Family: A Study of Work and Leisure in the London Region*. London: Routledge and Kegan Paul.

Zolotas, Xenophon (1981) *Economic Growth and Declining Social Welfare*. New York and London: New York University Press.

6 The question of goods (an indefinite epilogue)

If some limits are absolute, or nearly so, like the range of temperatures under which human life can be comfortably conducted on earth, or the tolerable level of toxicity in the soil, air and water; or debatable only within a narrow margin, such as the 350 to 400 parts per billion of global-warming gases that cannot be overtaken without risking run-away climate change, others are much more difficult to determine, or partly relative to subjective criteria: how many people can the earth sustain, together with (or as opposed to) how many other living beings, and of what kind? But even in the case of the hard, non-negotiable limits dictated by the laws of nature, the way we choose or are forced to adapt to them are related to personal or cultural values that are themselves relative. The earth can healthily sustain many more people on a vegetarian diet than on one favoring meat; even taking into account solely the preferences of the humans concerned, shall we be fewer happy meat-eaters or more numerous vegetarians, or fight over the privilege of who gets to be the former? Similar questions arise about access to, indeed the survival of, open space and wilderness, about the enjoyment of natural beauty, the pleasure of an unspoiled beach or a splash in an undammed river, the extravagant right to run a speedboat or a race car, or the simple luxury of owning a pet. In a way these are all, narrowly conceived, economic issues: necessary choices in the allocation of scarce resources, only made tighter because limits make scarcity more absolute. Limits restrict, as we have seen, the possibility of having it both ways, or all ways, by having more: the escape into growth. But in a larger sense, it is easy to also see that they make more explicit trade-offs that are often left unthought in economic transactions strictly conceived, choices between benefits and drawbacks, advantages and inconveniences, goods and bads, of very different kinds, forcing us to weigh one against the other "utilities," as economists would say, that are (to use an oxymoron) qualitatively incommensurate; to arbitrate, precisely, between (measurable) quantities and (inestimable) qualities.

The question of limits thus opens up what can also be called, using all the ambiguities of the term, the question of goods. Not only the material and countable goods usually embodied in commodities, and the less material but still measurable benefits provided by services (human or natural), but also those more or less tangible goods we variously value in and of themselves, sometimes above all else, according to our individual temperament, our culture, our social condition, or our metaphysical beliefs. The question of goods, in this loosely Aristotelian sense, calls our attention, beyond the things we value, to our values themselves, and ultimately to the necessary process of valuing, of choosing between different competing values that Aristotle identified with practical wisdom.[1] This is of course a practice that, by virtue of being human, we must always exercise, however reluctantly or even unconsciously, but the focus on limits makes it all the more explicit and deliberate: because we can no longer have all we want, we must decide what we really want.

This renewed consciousness, and acceptance, of having to make choices between goods that take into account ecological limits is most obvious – in varying degrees from the purely symbolic to the most radical – at the individual level, in the personal ethos and behavior of some of those of us in the affluent world who have the luxury of making such choices. It is perhaps best manifested in the diffuse movement known as "voluntary simplicity," but also in related currents favoring smallness (*Small is beautiful*, downsizing) and slowness (from slow food to localism), all deliberate refusals of the dominant trend toward more, bigger, and faster. Given the relatively minor impact these "lifestyle" choices have so far had on the consumption patterns of advanced economies, combined with legitimate suspicions of the superficial and self-indulgent motivations sometimes underlying them, it is easy enough to make light of their significance. And yet, seen in the perspective of a somewhat longer history of desire in the West, this (re)turn to self-limitation may signal the beginning of a shift as crucial as it is necessary. What it represents in practice, more or less knowingly or naively, but often quite consciously, is a reengagement with the tradition of wisdom that has in one form or another been cultivated by all civilizations at least until the early-modern era. The reflection on desire that recognizes that satisfaction does not always consist in giving in to it, and that greater fulfillment may be found in being content with what we have, rather than in the infinite pursuit of what we don't. And that, the possibility of an escape from the logic of lack, would constitute a momentous change in our relationship with limits.

For there is no doubt that a major shift in the history of desire has accompanied the development of capitalism in the West, and followed its gradual spread to the rest of the world. Although it is a history that should resist simplification, we can helpfully think of it as the liberation of desire that corresponds to, and is somehow the consequence or condition of the triumph of economic liberalism, for the freedom of trade ultimately requires the freedom of wants. As we already saw in Chapter 2, the story of this shift was most famously told by Albert O. Hirschman in his seminal *The Passions and the Interests*. Over the course of the long eighteenth century, the proddings of self-interest and of earthly desires (the passions) that had traditionally been viewed with suspicion as a threat to both spiritual and social harmony came to be regarded as not only harmless but beneficial to the growing prosperity of the new commercial economy (*le doux commerce*). Unlike the comparatively static order of the old agricultural regime, the economic dynamism of the fourth stage no longer required the economizing of desire, but on the contrary its infinite proliferation. It is in this sense that Jean-Claude Michéa in particular is so right to insist on the original complicity of social and economic liberalism,[2] so often obscured by later political realignments. Market capitalism is a libidinal economy. Rather than being by nature conservative, it is in the most literal sense *expensive*. As any participant-observer in its late globalized proliferation cannot help noticing, its most glaring manifestation is the immense enterprise of incitement to consume – the omnipresent mediatization of compulsory advertising – which is both a major constituent part of the economy itself and the clearest revelator/producer of its intended subject.[3] The modern, fully liberated subject of consumption would itself be consumed by a desire for ever more, under the spell of a limitless, because endlessly reproduced, promise of gratification.[4] Of course most of us still manage to resist the call at least part of the time, and to exercise the virtue of prudence in which Adam Smith, ironically, saw the more enduring human proclivity. But there is little doubt that the relentless exaltation of commercial fulfillment exerts a performative as well as normative pressure, urging us to embrace the consummation of material desire as the ultimate good.

Much the same can be said about the pervasive influence of a closely related concept: the fiction of an entirely self-interested *homo economicus*. Hardly any economist, least of all Smith himself with his complex understanding of human personality, ever thought that people were solely motivated by concern about themselves, and material concern at that. Rather, in seeking to give itself a scientific basis, the discipline of economics chose as a heuristics, precisely because there are so many other uncontrollable variables in human behavior, to reason *as if* human beings, on the average, pursued only their own material advantage.[5] It was a drastic simplification, originally assumed as such. But as economic thought, or economism, has become more and more hegemonic, and its premises increasingly naturalized, we have begun to forget the all-important "as if" clause, and may well be tempted to believe that human beings really are primarily self-interested, driven by a desire to "better their condition" understood only in material terms, and that greed is, if not good, at least normal. Now such an ideological shift – which is not so far from having reached acceptance that it is no longer shocking to liberal sensibilities – would not only represent a triumph of reductionism, but a betrayal of the wealth of knowledge accumulated from ancient times by so many traditions of moral reflection, both common and learned,[6] as well as of the newer understanding provided by the psychological and social investigations of well-being and happiness.[7] For surely we know, in our hearts of hearts and through the simple observation or scientific study of our neighbors, that if all of us can sometimes be selfish, and some pathologically so, human beings are not merely, and not even mostly motivated by material acquisitiveness and self-aggrandizement, but just as much by our need for the recognition (what Adam Smith called the approbation) of others, and for their affection, not to mention the considerable pleasures of disinterestedness. And that money, in the end, really can't buy us love.

The acceptance of limits, enjoining us to free ourselves from the logic of growth, inevitably leads to the rehabilitation and reinvention (when it is not a complete rediscovery) of the virtues of moral restraint and self-cultivation that have been so massively disparaged as moralistic, and perhaps even more efficiently, discredited as puritanical, by the uniquely modern combination – far from always conscious and assumed on the part of the millions of us who fell prey to it – of libertarianism and consumerism which is the natural ideology of market capitalism. It is probably no mere coincidence that among the pioneers of environmentalism who always insisted on the dimension of personal ethics as well as on the economic challenge of a transition to a sustainable life, so many acknowledge, more or less discreetly or publicly, the grounding of their commitment in some religious tradition, or less explicitly in a form of spiritual practice.[8] It takes an especially strong moral compass to be able to resist the ambient siren calls to material consumption, and one that is in some way exerting a counter-influence to the overwhelming current of modernity, or has at least remained relatively immune to it. It is therefore natural that those individuals who still feel connected to one or more of the established traditions of wisdom and self-restraint inherited from the past or being re-invented by contemporary religions would find themselves better equipped, and indeed find it more fulfilling, to live in a less material world.

Which does not mean, of course, that it is sufficient to belong to a religious tradition to become a good environmentalist. It is on the contrary a widely shared and persuasively argued opinion that the great monotheistic religions, with their emphasis on the supremacy of man as the apex of earthly creation, and their proclamation of his complete dominion (under God) over all other living and non-living things created solely for his own benefit, not to mention the well-known injunction to be fruitful and multiply, have provided ideological support and theological justification to our instrumental and exploitative relation to the world.[9] Certainly, without going so far as adducing direct historical causality, there is a strong association between the birth of theistic religions and the development in different parts of the world of the increasingly monumental, hierarchical, and frequently destructive civilizations (including of their own environment) that agriculture first made possible.[10] And jumping to the other end of this long and entangled history (*pace* Feuerbach and Max Weber), we still find that among the staunchest deniers of global warming Christian evangelists stand out prominently, and not only because the most devout among them really would not mind hastening Armageddon.

But on the other hand, the very same religious traditions, or other currents within them, can equally lead to an emphasis on the custodial or stewardship role entrusted to human beings by virtue of their superior status. Precisely because we are divinely endowed with a higher form of consciousness and unparalleled abilities, we have the unique responsibility to preserve and nurture the rest of creation for the benefit of future generations, much as Karl Marx believed.[11] Contrary to what the most uncompromising anti-humanist principles of deep ecology seem to imply, an insistence on the privileged position of human beings in the universe, divinely ordered or not, is not incompatible with the makings of a good environmentalist: after all, we would all have to agree that only we, the humans, have a choice. But neither, alas, does it suffice. The more nuanced truth is that established religions, inheritors of the long and often conflictual histories that have necessarily taken them well beyond their foundational sources and into the present, offer, in this realm as in others, a range of interpretations more or less conducive to ecological consciousness. As in the case of the debate between the believers in free-market solutions to the environmental crisis (green capitalism) and those who see capitalism as the problem, we will all be better served by adopting the more embracing, and yes, catholic attitude. If people of faith find in their own religious traditions, or in others (witness the powerful appeal of Eastern spirituality for the many who are disillusioned with materialism in the West) support for their environmental commitment, it is undoubtedly a good thing, and if this faith has inspired them to become leaders and shining examples in the movement to restore the earth, we should be grateful for it.

Yet this still leaves out the countless lost sheep in the world for whom such appeal to religion or transcendence is not a viable option, or whose personal beliefs give them no particular direction. Will there be for them no more persuasive resource than the cold facts of science, and should this in fact be the only reliable motivation? Freeman Dyson (himself a scientist) makes the interesting suggestion that such an insistence on science and economics misses the main point, which is that environmentalism has itself become

> a worldwide secular religion [...] holding that we are stewards of the earth, that despoiling the planet with waste products of our luxurious living is a sin, and that the path to righteousness is to live as frugally as possible.
>
> (Dyson, 2008)

To be sure, one can have serious questions as to whether such tenets are sufficient to make of environmentalism a religion according to most traditional criteria. After all it so far has no priests nor hierarchy, no temples, no sacred founding text (at best some highly regarded meditations and prophecies), and no gods. The last point may in fact be its greatest ultimate strength – it would be an immanent religion, at most a kind of diffused pantheism. But Mr. Dyson (and all the other fellow travelers who must have had a similar intuition) is surely right to insist on the moral dimension of the movement, even and especially when he uses, perhaps with some ironic distance, such loaded words as sin and righteousness. Environmentalism would be a religion – a re-linkage between all of us – that taps into our deep ethical impulses and longings, even, why not, our long overdue sense of guilt toward the earth (so often, need it be said, imagined as maternal), and hope for attunement. Whether this is "a religion we can all share," and whether, as Freeman Dyson further elaborates in the same tone, "[s]cientists and economists can agree with Buddhist monks and Christian nativists that ruthless destruction of natural habitat is evil and careful preservation of birds and butterflies is good," remains to be seen. But the need for a shared environmental ethos, for a motivation to care for our world more compelling than rational and scientific arguments have so far proved to be, for a moral commitment to values beyond ourselves, yet not necessarily transcendental, an ecological morality that more and more of us could embrace in spite of our differences – this much at least seems convincing.

Still, individual convictions, essential as they are, will never be enough. Not only in the more obvious, practical sense that the sum of our personal decisions and engagements, however intense, cannot have sufficient weight in the real world of institutionalized powers and vested interests to affect the necessary change if they remain individualized. This is the formidable political question, made all the more intractable because the world of politics is still largely organized on the basis of competing sovereign nations, an organization so ill-suited to resolve a global problem wholly oblivious of national boundaries. But individual resolutions are also inadequate in the only slightly less obvious, and somehow inverse sense, that when it comes to values, we are hardly individual. Rather, the convictions we hold dearest and most personal to us as individuals are rarely separate from those of the larger cultures and histories we are heir to, even when we think we are rebelling against them. They are greater, more ancient, more persistent and immune to willful decisions than we are, even though – this is the wonder and mystery of culture – they do not exist outside of each individual consciousness (and for historical civilizations, in the monuments and documents bequeathed from past individuals). This does not mean that cultural values do not change – in our hypermodernity we may even have the impression that some are changing very rapidly – or that they are impervious to individual influence, particularly when it is organized in collective movements. But traditional systems of values and cultural dispositions – what used to be called civilizations – have their own logic and rhythm, and their own resilience. This is bound to have considerable bearing on the question of limits we must all confront together, and the question of goods we may not share.

Admittedly this is a difficult subject to address without succumbing to various pitfalls, unwarranted generalizations and the lure of prejudice being the most evident. Or worse: for a long time, the study of civilizational values was uncomfortably close to forms of racialist thinking. Largely for this reason, modern scholarship has tended to stay away from the comparative study of ethical orientations throughout the world (apart from the study of religions themselves, which is not necessarily the same, and even then is perceived as fraught with danger). Of course, cultural anthropologists still study value systems at the micro-level, but not since the great American culturalist school of the mid-twentieth century, particularly with the work of Ruth Benedict and Margaret Mead,[12] have serious thinkers dared to compare ethical cultures on such a large scale. When contemporary scholars engage in this perilous enterprise, such as sociologist Emmanuel Todd in his long-term study of the influence of traditional family structures on ethical orientation,[13] they are at best viewed (from a distance) with grudging consideration, and more often with suspicion. These are reasons enough to be cautious. And perhaps in the globalizing twenty-first century, with its great international flow of merchandise and information, let alone of populations, economic forces and the distribution of resources will be much more determinant than those cultural differences that manage to resist entropy (Claude Lévi-Strauss (1997, 472), at the end of *Tristes tropiques*, proposed that anthropology should give way to entropology). Yet it may be unwise to discount cultural goods entirely.

A prominent case of the importance of cultural attitudes for the future of our planet – some environmentalists would plausibly argue that it subsumes the whole field since it concerns the ultimate good – is how, how variously, and how much, we value nature. Nature in all its aspects and conceptions: the natural world, untamed or transformed by us, the visible and the invisible, life and unlife, oceans and their whales, mountains and microorganisms, our little infinite share of the cosmos. Nature in its many meanings – and perhaps meaninglessness – to many people. What is it worth to those who make their home and living in the Amazon, in Hong Kong, Eritrea, or Montana? Now science, and economics too, have their own objectifying answers to the question of the value of the natural world. Environmental scientists everywhere are laudably and often desperately trying to make us see how truly essential for the equilibrium of our host planet are the last rainforests and tidal marshlands, the proper acidity of ocean waters and the replenishment of aquifers, the albedo effect of glaciers over land and sea, and the multiplicity of species. Economists too are beginning to put a price of these goods, and we had better listen to them, for it is very expensive. But in a sense this assessment of value, fundamental as it is, remains instrumental. In the (purely hypothetical and in fact impossible) event that we found a way to replace the rainforest by catalytic converters, the oceans by purifying (industrial) plants, glaciers by giant parasols, and to synthetize new species as fast as the old ones are vanishing, everything might be fine from an objective point of view. And yet many of us can see that in another sense of value, the world would be radically impoverished, and for some, hardly worth living in. How important is this other good, this other appreciation of nature, for different people, and peoples?

In the history of the West the changes and particular forms taken by the sentiment of nature as a source of beauty, solace, inspiration or wisdom have long been studied and interpreted in great detail, with their moments of efflorescence in Romanticism and American Transcendentalism, and their merging with the new ecological consciousness and movements for the preservation of (not too spoiled) natural environments in the twentieth century. In the East too an even older history of exquisite writing and painting, as well as numerous traditions of sages finding enlightenment in the solitude of the forest, testify to a deep appreciation of nature for its own sake, at least by those who could devote themselves, as in the West, to the cultivation of its privileged manifestations, poetry, the landscape, and contemplation. But what about others? Is such a sentiment for nature specific to, and in fact a privilege of those particular civilizations?

There is, a priori, no reason to think that the mother who picks up the few remaining twigs at the edge of the Sahel because she has no other fuel to boil her precious water, or the displaced villager cutting down a last stand of primeval forest in Borneo to supply Moondeer with (hopefully fair trade) coffee, do not occasionally share a moment of wonder and sorrow at the sight of their vanishing world. For those peoples who did not produce pictures of their natural surroundings or elegies to trees, but perhaps identified with them so closely that it was not their form of homage to represent them, the history of nature, art or even philosophy may have little to say, but how great is their loss? What we do know is that they, like the Asian descendants of those who made exquisite ink drawings of serene ponds and cerulean mountain skies, and of those who celebrated that other pond in still relatively countrified New England, have more or less learned to survive in some of the biggest cities in the world, and the most wretched, amid the least natural ugliness. At what cost such dispossession, not even alleviated by the artificial beauty and comfort found in many Western cities? But of course, we should not forget that in the West itself the different waves of romanticism were exactly contemporary with those of the industrial revolution, with its own brand of ugliness and brutality, and no doubt a compensatory discovery of the good that was being lost in the new splitting (but not of course the first, which is as ancient as Gilgamesh) between humans and the natural world.

And even within the West there have historically and locally been very different ways of coping with this loss, of distributing it as it were, and of preserving some of what was lost. In long settled and cultivated Europe, with its much older and denser population of agriculturalists, very few natural areas, except in high mountain ranges, have been left untamed and uninhabited, and none in a condition of original wildness. But almost everywhere, even in the most industrialized countries, sufficiently prudent use and attentive occupation of the land have created and preserved that delicate balance between natural landscape and cultivated land called the countryside. The countryside is real country. It is still a valuable and productive agricultural resource, but for most people all over Europe, and many outside of it too, not only for those who live there permanently but especially for the dwellers of cities, it has another inestimable value: as fresh pastoral air, peaceful refuge from modernity, reminder of history and often of their roots, and as what they think of as nature. Thus the countryside in the old world is worth cultivating for its own non-instrumental sake, even if it is less productive than the most extensive land use, even if it is not fully rational economically. It is a natural and a cultural good.

In North America, except for those parts of the East and the South that were settled in colonial times on a European model, from the Mid-West to the Central Valley of California, there is no countryside. There is agricultural land of course, some of the most productive in the world, but it is industrialized, heavily mechanized agriculture, in a sparsely populated rationalized landscape where nobody goes for pleasure. Where people do go for pleasure, from all over the world, are the famous national and state parks, and many regional ones besides, preserves of that immense natural treasure called the wilderness (to complete the picture, one would have to add the millions of acres of extensive ranchland so iconic of the Western landscape, neither completely wild nor domesticated, but unlike any European countryside). So Americans have made a different choice. Because their much later immigrant expansion and lower population density had left vast expanses of nature largely unspoiled, and because a few environmental pioneers had the vision to see the incomparable value of this heritage before it was too late, they convinced their fellow-citizens and their presidents to set it aside for ever and for us to enjoy. But they gave up on all the rest, gave up on the ecological health and cultural benefits of farmland for purely instrumental exploitation. In terms of environmentalism, it is close to an all or nothing policy,[14] whereas the Old World approach is closer (but only relatively so) to the integration of nature and culture once imagined by nineteenth-century utopians. In each case something has been lost, but a different natural/cultural good, or a distribution of the natural and cultural emphasizing different goods, has been preserved, because people in North America and in Europe still hold some portion or version of the natural world to be of higher value than the benefits of further economic expansion, or they have not yet had to face on their own land the ultimate trade-off between what are, in the end, irreconcilable goods.

But will it always be so? For one can have diverse opinions about the merits and relative sustainability of each cultural model, or a preference for other possible ways of preserving natural goods, but the final question is whether, and for how long, these goods can sustain the pressure of continued growth. Will they eventually give in to it, or turn out to be strong enough to convince at least the affluent West, old and new, to renounce growth before it is too late? And more urgently, can these goods even be afforded by the rest of the world, where under any equitable assumption growth must continue for the foreseeable future?

One thinks of South America, with the largest extent of primeval forest left on the planet and its incomparable depth of life and biodiversity, the veritable lungs of the earth, and only possible home to some of the last people still living at the first stage today. Here the natural goods are inestimable and esteemed by everyone, even one supposes by those who are directly involved in destroying them, yet the human pressure is too great to stop, and unlikely to diminish for a long time. How small will the Amazon International Park eventually be, if it ever comes into existence? Hopefully large enough to sustain a museum, and perhaps a zoo, with the descendants of the present-day hunters and gatherers preferably serving as keepers rather than inmates. And if this seems too grim a prophesy, one should think of Africa, where natives do keep the dwindling wonders of their national parks from the desperate predation of fellow natives for the benefit of first world tourists (truthfully for the benefit of all of us who will never have a chance to get there, but know at least that such wonders still exist). And this is the continent that on average has not even begun to grow at rates resembling those of the rest of the world, and is so far from any prospect of catching up that it must develop exponentially, or for a very long time, if justice is ever to be achieved among humans here on earth. Does this mean that unspoiled nature has become a luxury good, a privilege of the rich countries that can still keep up and protect their pleasure gardens, while the rest will go the way of China's gorgeous gorges, sacrificed to the very real needs of its aspiring population?

A particular subset of the natural world that receives an exceptional level of attention, appropriately though not always with the most justifiable focus, is that of the creatures now frequently referred to as non-human animals (thus recognizing our kinship with them at the same time as denying their kinship with us). And there is no doubt that our relationship with animals has long been one of the most symbolically and emotionally charged, even if contradictory – thereby qualifying them as a highly invested good in our value system – as well as practically and economically consequential since we turned them into commodified goods, and not coincidentally into an environmental menace. For in this domain also we have divided the natural realm between the wild and the tame, between the virtually untouched and the thoroughly domesticated, even thoroughly bred for our utilitarian purposes, but with very little in between (except perhaps for the domestic cat) so that we have developed an almost schizoid relation to animals – in the original sense that we have ourselves lapsed into a divided and incoherent subjectivity, if not multiple personalities. The last time humans related honestly (though evidently not without conflict) with animals was when we still subsisted entirely as hunter-gatherers, killing for ourselves the wild beasts we consumed. Since domestication we have been of at least two minds, and remain divided, from children picture books and baby foods to wildlife documentaries and nursing home dinners. And so with regards to the questions of goods and limits there are now two completely different animal realms and different problems – except for the common dilemma of how many of them versus how many of us there are to be.

As a good, the wild animals present a striking paradox. The ones that attract the most attention and are apparently the most valued emotionally (leaving aside the lowly worms for the moment): the lions and tigers and polar bears and elephants and pandas, and the many other spectacular ones that are the stars of the documentary film industry, may well from an objective and ecological point of view have the least value of all. Of course in their natural environment they play an important part in shaping the landscape and using resources in a particular way, and since many of them are top predators, we know that their removal would greatly alter the balance between grazing animals and plant foods (as it has in the settled lands from which most great cats and wolves have been eliminated) so that their former prey would have to be controlled some other way, most likely by replacing them as is everywhere the pattern by domesticated animals. The eventuality of their disappearance, therefore, would not be without consequences (including good ones from an economic standpoint), and given the degree of unpredictability always involved in ecological disruption, it is not to be taken lightly. Still it could be coldly argued that if all the most prized wildlife on the planet were eliminated, or parked in zoos and reserves as is increasingly the case, it would not endanger our environmental survival in any serious way, certainly not in comparison with the ravage we are currently inflicting without much remorse. From this perspective, it would appear that the top-tier wild animals, the prestigious ones, are a case of purely symbolic good. Their only value is as aesthetic or emotional objects for us, for the pleasure, the admiration, even the wonder they make us feel. Of course, this is not true from a non-humanistic or a theological point of view that would consider all life-forms of equal worth or sacredness, for example, but from a strictly human perspective, they are only as precious as we want them to be, their sole utility being subjective, which makes them vulnerable.

This precarious position of the wild beasts as a purely subjective good (perhaps in common with that of their cosy cousins our domesticated pets) is not in itself sufficient to know what their fate will eventually become in a limited world. They are after all also a luxury good, and as long as a high level of inequality and dispensable wealth prevails their preservation in something resembling their natural environment may remain economically attractive along with the rest of the luxury industry.[15] Or alternatively, in a more ideal world, the international community as a whole would keep embracing them as the irreplaceable natural/human good that they are and commit enough resources for their survival in a suitable habitat. But this good has to be weighed against the fact that the last remaining wild animal populations are predominantly located in the poorest regions in the world and/or those with the greatest demographic pressure, mostly in Africa and Asia. How much can their preservation, and that of the vast territories they require, be worth to the human populations for whom they are often an obstacle to development, with whom they compete for already scarce resources? What is the unaffordable price of natural beauty when you are starving, or simply aspiring to live a little bit more like the average person in the rest of the world, from which wildlife was eliminated long ago? In this game of how many of them versus how many of us, the odds are not in their favor, though their fate remains entirely in our hands.

The fate of their domesticated non-human kin is equally in our hands, but it involves very different goods, and it is so far on an opposite course. If the wild ones are likely to see their numbers indefinitely reduced and end up in zoos and reservations (where they can no longer live their natural life but are at least treated as humanely as possible), the vast majority of their much more numerous domesticated brethren are already living in conditions much worse than zoos and reserves, with little concern for their well-being, but their numbers are likely to keep growing (if only for shorter and shorter life spans), for unlike the former they have enormous economic value as commodities in increasing demand. They also have a huge environmental impact, in many respects nothing short of catastrophic.[16] If only for these few reasons, their proliferation should be a major concern in a world of limits. In the realm of goods, they are especially bad.

Not that these issues leave everyone among the humans indifferent. In the field of animal studies and of environmental ethics generally there has long been considerable attention to the way so-called farm animals are treated, to our rights and obligations toward them, and the justifications for our instrumentalization of their life, only second to the perhaps more difficult ethical question of laboratory animals experiments for (human) medical research.[17] From ethology, anthropology and the science of evolution, we are inexorably (if defensively) learning that they are much closer to us than we once thought. And at the very least, few persons of good faith would now dispute the answer to Jeremy Bentham's (and later Peter Singer's) trenchant question: it is not "Can they *reason*? nor, Can they *talk*? but, Can they *suffer*?"[18] As many of Descartes' contemporaries already knew, they do suffer.[19]

How much? It obviously depends on the conditions in which they are "farmed," ranging all the way from tolerable for those that roam and graze with a degree of freedom for most of their lives until close to the end, to abominable for the much greater number confined and compressed from (artificial) conception in breeding and milking or fattening factories. Not to mention the slaughterhouses, often much rationalized since the days of Sinclair Lewis' *The Jungle*, but for all their efficiency still industrialized torture chambers. This massive abuse of indisputably sentient and suffering non-human beings is well documented by animal activists often derided as radical (the more mainstream organizations seem more concerned with the pain of neglected pets, or too polite to confront the meat industry) and for everyone to see who would have the heart to. And there is little doubt that almost anyone who was exposed to the reality of living-meat processing would feel queasy at least for a little while about continuing to participate in it. Or if they didn't, like the badly paid workers who forcibly become insensitive enough to do the dirty job for us, it would only prove the insight already arrived at by Thomas Aquinas and Immanuel Kant,[20] if not already in Antiquity:[21] when we get inured to the cruel treatment of other sentient beings, we get used to cruelty itself, including to our own kind. In behaving inhumanely toward non-human animals, we risk dehumanizing ourselves. But of course, most of us urbanized moderns avoid this risk by staying well away from the scene of the crime. Unlike our more honest ancestors not so long ago, and the many current pastoralists in the rural world who still butcher their own meat or routinely witness the slaughter of familiar animals, we protect our humane sentiments by distancing ourselves from the suffering that nourishes us, by dividing ourselves from our own reality.

Are such sentimental qualms then another luxury of the affluent West, with little consequence for practical prospects? For all the genuine commitment to animal welfare among some subsections of Western populations (whether young and idealistic or old and more thoughtful), and some lessening of meat consumption in a small conscious minority (though often also for reasons of their own health), the richer countries remain carnivorous ogres,[22] and can hardly serve as models or lessons for the rest of the world. In this respect several other cultures have a much deeper track record, whether they have long refrained from eating flesh out of traditional belief or based on religious prohibition from taking life or inflicting suffering on sensitive creatures, and their experience offers more enduring examples of a different way for humans to share the world with our fellow animals. Or at least proof, if it were still needed, that it can really be done. But then these are still not reasons enough to do it, or likely to convince everyone that it is a better way, a higher good. Other cultures will undoubtedly invoke their own traditions of hunting, or ritual sacrifice, their belief in the divinely ordained right of humans to use other creatures as they please, or simply their overriding penchant for carnivorousness, because after all, meat is good.

It is also true that even among those who are already convinced, or can be swayed, by appeals to the sensitivity and sufferings of fellow creatures, not many will extend those protective feelings beyond those animals that seem to feel and suffer like us, with whose life we can identify. Few, unless obeying an overarching sacred prohibition against destroying any life, like the Jain, will be terribly moved by the sufferings of crabs and prawns, or fish for that matter, or even, let us admit it, rather stupid chickens. But these lesser creatures from a humanistic point of view are just as important for the damage their use and misuse by humans can inflict on the planet, both in the seas and on land. From the point of view of limits, they matter as much, perhaps more, which is why another set of arguments, more properly ecological, and eventually economic, must also be brought to bear in addition to our anthropocentric capacity for empathy with other animals.

Until fairly recently, we were principally agriculturalists and animal farmers only on land. In the seas and oceans, even rivers and lakes, we were still hunters and gatherers: we fished and harvested wildlife, largely left to its reproductive and growing patterns in its natural habitat. It is no longer the case. Not only have we emptied out the oceans of entire species of fish and brought many others to the brink of healthy survival by overfishing them on an industrial scale, ravaging in the process the sea bottoms and their precious life sources by the use of giant trawlers – this is the damage caused by our modern hunting and gathering itself, in addition to that caused by our poisoning of ocean waters with pollutions of all kinds, creating quasi continents of plastic, extended dead zones due to eutrophication, and raising acidification and temperature levels to dangerous heights.[23] But by now, in compensation for this excessive emptying out, we have increasingly taken to breeding and growing these fish and other sea animals on land, as it were, or by the seashore, in so-called "farms," so much so that the global consumption of such (agri-)cultivated seafood has overtaken that harvested by all fisheries combined, with the devastating consequences of all industrial farming for the environment.[24]

For these giant concentrations of thousands of live animals in confined spaces, cages, pens or tanks, whether they be pigs or cattle, fowl or fish, or the ever-multiplying prawn, are unnatural disasters. For the animals themselves of course, deprived of all semblance of their original habitat, but also for ours. Such concentrations of organic life necessarily produce masses of waste and effluents that must be disposed of and somehow prevented from leaking into the ground and water. Because they also offer ideal conditions for disease and contamination, the confined bodies therein must be relentlessly treated with antiseptics that find themselves in the meat we eat or join with the other toxic by-products seeping into the environment. Not to mention the irresistibly comical but not at all funny fact that the methane "emissions" of all those farmed cattle are now second only to those of fossil fuel burning as a source of global warming gases.[25] And more fundamentally, because the animals in such concentrations cannot have access to the sources on which they would naturally feed, they are fattened artificially, and increasingly unnaturally, thus contributing to the expansion of the worst forms of industrial agriculture. They account for an enormous proportion of two of the most extensively mechanized, fertilized, chemically treated and modified crops in the world, corn (maize) and soya beans – and these, at least, are plants, somewhat more natural to herbivores than the meat by-product protein flours fed to normally grazing animals, thus effectively turning them into self-cannibalizing carnivores. Or to carnivorous fish and the voracious shrimp, the so-called trash-fish now massively harvested to supply the ever-expanding seafood farms, thus further depleting wild ocean populations.[26]

And finally, if this is not enough, the economic argument: feeding animals for human consumption is incredibly inefficient; depending on the kind of animal grown, it takes between three (for chicken) and seven times (for beef) more plant production to feed us on a meat diet than it does if we eat plants directly. The proportion of fossil-fuel calories we must expend to produce the equivalent in food calories ranges from 4/1 for chicken to 40/1 for beef and 57/1 for lamb, and producing animal protein requires about 100 times more water than the same amount of grain protein. By any ecological or economic criterion we look at it, breeding animals for food is a disastrous proposition.[27]

So this lesson should be clear. In the case of domesticated animals that exist solely for our consumption (pets represent for most people a very different good, which does not mean they are not a luxury), there should be much fewer of them in proportion to the number of us. And, remembering *Diet for A Small Planet* (Moore Lappé, 1971), we should not eat meat. Or if we really want to, because it is a good we value so highly, we should do so rarely, giving proper thanks, and from animals that have had the best possible life, preferably grazing freely in areas that are not suitable for any other use (and given the urgency and magnitude of the need to reabsorb greenhouse gases, this should include reforestation). Or where they are still naturally abundant in the wild, those of us who have the heart can hunt them; that is still the most honest way.

Do these conclusions, so often reiterated by almost everyone who has cared to look into the issue, have any chance of being heeded? At the moment, the trend is in the opposite direction: global meat consumption is increasing, together with the pressure to produce it more and more cheaply, in worse and worse conditions. And this is not surprising. For most of even recent history, most of the people in most of the world did not eat much meat, not because they did not want to – humans have a long anthropological record of privileging meat even when they do not actually consume much – but because they were too poor. As more and more of the world escapes from poverty into relative affluence (the obviously good side of growth) many naturally desire to satisfy their craving for more animal food. And this may be particularly pronounced if the faster and larger developing countries happen to have a strongly carnivorous or piscivorous culture, as some, though not all, currently do. This would point to severe conflicts over these particular goods, with the inevitable question of equity their distribution raises. Why shouldn't the rest of the world aspire to become as unsustainably carnivorous as the affluent West? To the sardonic slant of this rhetorical question, there may be one slightly hopeful rejoinder. It is possible that the evolution of meat consumption within a developing society (regardless of culture?) follows a Kuznets curve: as meat-deprived people first get richer, it starts rising sharply and bells up to a high level, but then as they settle into satiety, and perhaps become more health conscious or even concerned with the welfare of non-human animals, the curve tapers off, and might eventually go down, as is apparently happening in some privileged segments of Western societies. This is a very optimistic scenario; it would take a long time to run its course on a global level, and scarcity will probably decide the issue sooner and more crudely, but it would be a good resolution to this question of goods.

There are therefore many reasons to focus on the fate of the most familiar and visible animals (though they are often out of sight). It is easy to see how it involves values and interests – economic, ecological, moral and cultural – between which there are real conflicts, at once complex and concrete, that are nevertheless in our power to confront, if not to resolve easily. Even among the other life forms with which we share our planet, however, they may not be the most important, or present the most pressing problem. Far more crucial are the myriad invisible ones, and all those lowly worms we are barely conscious of (except for a few dedicated specialists) and yet are the foundation of life on earth itself. In their case it is not a question of the numbers of individuals being threatened, or even of some species on the verge of extinction, but of entire *genera* being eradicated every year as a (usually involuntary and undetected) effect of our activities. How many and with what long-term consequences? We often don't know because many of these creatures had yet to be discovered, let alone studied, before they disappeared forever.[28]

So what is the value of this good, of this irremediable loss? Here the most ingenious efforts of prospective economists, even the best estimates of the most informed scientists, must fail us. We are in the realm of the truly immeasurable, though nevertheless absolutely certain (more certain indeed than the future effects of global warming, about which there is increasing though so far largely ineffectual alarm). One would therefore think that it would rank very high, if not highest in the concerns of many people with different, even opposite value systems. For those who believe in actual creation by a supreme being to whom we owe supreme respect, this genecide through our own works should truly be seen as the unraveling of genesis,[29] the utmost disrespect for the divine Works themselves. But even for the pure Darwinists who think that everything is merely the result of the selective survival of random mutations, and that the disappearance of some species, after all, has always been part of the natural process, it should remain that all those species we are destroying in a matter of decades or centuries had taken millions of years to evolve, and that it would again take (if *we* disappeared) millions of years for new ones to replace them. By our perception, it is a slow and almost invisible erosion; in the timescale of the universe, it is an incremental apocalypse, happening now, not in some distant future (though it is likely to accelerate), and entirely of our own undoing.

For the cause, in this case even more than in the previous ones, is painfully clear even if it is more diffuse: there are too many of us, consuming too many things. Of course in some particular situations, if sufficiently understood and soon enough, specific remedial measures can be taken and some species saved: a unique frog in this marsh, a wondrously adapted alga in that brackish current, thanks to the patient dedication of humans who understand their value. But these heroic efforts to preserve patches of earth cannot hope to stop the crushing advance of our collective footprint, not as long as there are ever more feet on our limited ground, and their shoe-size keeps getting larger. So in the end it is demographic and economic growth; the growth of human population times the growth of human consumption. What good are those, against the loss of all the other goods? That is the question of limits.

As to population – the too many of us – we have seen that an increasing majority of people all over the world no longer consider greater numbers, as most did in the past, to be a good in itself. Apart from the very poor in regions that have not yet fully engaged in the transition to modernity, and some fundamentalists who still think that their own tribe should be fruitful and multiply at the expense of others, the average human now favors having no more children than will insure her own reproduction, and in many affluent countries, fewer than the reproduction rate. The quality of human life – that of the children and of their mothers – has become more important than the number of human lives. This is a profound revolution in the history of goods, and also one with immense practical consequences; it is this crucial change in mentalities and behavior that allows us to hope for a stabilization, and then perhaps a gradual decrease, of the total human population inhabiting the earth. But of course, demographics, like climate change, have their own momentum. Even though our collective fertility is currently decreasing at a spectacular rate compared to modern historical trends (a development much more hopeful than in the case of global warming), the impact of this virtuous cycle on total world population will not be measurable before the next several generations (were we equally virtuous in reducing global warming emissions, it would take even longer), and so no matter what we do now, the number of humans on our limited earth will keep growing, according to most projections by at least another two billion.[30] This factor of growth, good or bad, is unavoidable.

This leaves consumption. The matter and energy (but not the time) each of us uses in doing the things we do, depending on how we live, how much we own, what kinds of buildings we occupy, how we keep them warm or cool (depending on where we insist on living), how and how much we travel, how far and how fast we want things to move about, what we do for leisure and with which toys (those power boats on the lake, or do we prefer to swim). All these activities that embody a certain degree of materiality, consume some earthly goods in the ordinary sense of commodities, and therefore contribute variously to our ecological footprint, also represent for each of us very different goods in the moral sense (utilities for economists), more or less fundamental or frivolous, indispensable, or not really, to our conception of a meaningful life.[31]

Confronted with the need to reduce our global footprint – a necessity much of the world is still trying to deny, but that the reflection on the question of limits forces us to accept – the different question that must therefore arise, or rise again, since it is nearly forgotten, is how much we are truly attached to these goods. Are we so wedded to them through our materialism, so deeply dependent on that version of the good life, or in fact quite capable of giving up some of it without a great sense of loss, perhaps even with a sense of liberation?[32]

This question, of course, is most urgently addressed to those of us who already have and consume the most, in those regions of the world, the First world, that are very largely responsible for overstepping global ecological limits with their outsize footprint. The others, those who are too poor to have to renounce anything, and the aspiring and ascending ones who have not yet come to identify the good life with the consumption of goods, are perhaps better equipped to accept the virtue of material limitation. They may still be closer to cultural traditions that, out of necessity or choice, have always privileged spirit, or affection, over matter, and detachment from wealth rather the infinite pursuit of it. Instead of encouraging the rest of the world, still by far the majority, to continue being attracted to the lure of endless material emulation displayed by the West, and expect everyone to join with the most affluent nations in a catch-up race our common planet cannot support, it would be better, it is indeed high time, to begin reversing the direction of modeling, or at least seeking a more balanced model. Let those who are still living within sustainable means become the ideal to which the world aspires, and provide the profligate nations with examples of the wisdom and moderation, indeed of the acceptance of limits, that seem to have been forgotten by modernity (which does not preclude, of course, their benefiting from the modern technologies that can help ensure sustainability).

But it is not certain that even the affluent West itself, apparently so committed to a culture of commoditization and material aggrandizement, has really left those traditions so far behind – not at any rate as far as the hyper-mediatization of consuming desire that has become its ideological screen would have us believe. As we already noted on several occasions, when surveyed as to what contributes most to their wellbeing and happiness, people in the richest countries consistently say that beyond a moderate level of income not far from average, it is not material wealth that makes a difference to their satisfaction in life, but how they *are in relation to* other people.[33] Once they feel comfortable enough (and that is already in itself a relative perception), what matters most to most people is not to have more goods, but to have the recognition of their peers.[34] To be recognized as peers, that is to say, as equals.

And so in the end, as we have come to recognize at so many turns in this reflection, the good we find most insistently affirmed at the heart of the question of ecological limits is a profoundly human one, as thoroughly moral as it is political: the need for more equality among us, or to put it the way it probably affects us most keenly, the need for less inequality. Given the complexity and contentiousness of the debate over the value of equality, and how difficult it has so far proved for our species to achieve lasting progress toward a more equal status among its members, both within and between societies, asserting the centrality of human equality – an extraneous good, after all, something of an externality – to the problem of environmental limits may well appear to be a disastrous strategy, almost guaranteed to be self-defeating. And perhaps it will be. Except that it is not a strategy at all, and neither is the good of equality external to the possibility of resolving the ecological crisis, but integral to it, and our best hope yet. And this, not so incidentally, is largely the reason why, while a genuine concern for the environment is not in itself, or should not be, a preserve of either side of the traditional political divide, political ecology has always found itself drawn to the Left, precisely for its greater attachment to the moral good of equality. Only on the condition of lessening the gap between rich and poor, especially by lowering the extravagant consumption levels of the privileged, those that could not possibly be generalized to the whole of humanity, can the idea of limits be accepted by a vast majority, embraced as an equitable goal, and finally valued as a good in itself.

Equality – at any rate a much more perfect equality – is the *sine qua non* of our living in ecological balance, the very hinge of a steady-state world. If inequality has indeed found an alibi in the imperative of economic growth, or to be as generous as possible, if it can be said – as its proponents do argue – to be justified by the emulation it provides for the competitive pursuit of growth (an argument largely belied by the most successful of the competitors, who generally profess to be motivated by other goods than material success itself), such justification disappears completely in a stationary economy where the ultimate good is for everyone to live within sustainable limits. Then inequality becomes not only a disincentive to pursue the good life, but an inherent injustice. And on the other hand, the more equality there will be among us, the more we will be at peace with the necessity of limits, and indeed able to recognize in equality one of the greatest goods of living within limits.

It is true that even if we achieve such a human and ecological balance, not everyone will be equally satisfied. One of the sobering lessons of the great culturalist school of anthropology evoked at the beginning of this chapter is that no one societal model can make everybody happy. In a more equal community valuing a life of contentment within accepted limits, competitive personalities who now thrive in the most aggressively unequal societies, the Randians and Wall Street wolves, would likely feel frustrated in their ambitions and insufficiently rewarded, and sometimes develop sociopathic behaviors, as the warrior types (who would have done well in many ancient heroic cultures) unfortunately do now in our more peaceful (but still competitive) societies. But we should remember that those exceptionally successful Alpha humans, predominantly but no longer exclusively male, who currently reap most of the rewards, symbolic as well as material, in our modern celebrations of inequality, do so at the expense of the vast majority. Inequality has a very high social cost, and studies that have attempted to measure it show an impressive correlation between the level of inequality and a large range of social and psychological ailments: physical and mental illness, murder and suicide rates, substance abuse, depression, and simple unhappiness.[35] And these findings should not surprise us. In a highly competitive and unequal culture where winners take most, and where the standards of material success reflected by the omnipresent media are attainable by only a few, most of the rest end up feeling like losers, or alienated and marginalized, or at best finding vicarious compensation in the imaginary of the star system. A society aiming for the more equal balance appropriate to a limited world may not be as scintillating, and offer as much excitement, or as many illusions, to the restless, and it would certainly not lavish such material extravagance on its high achievers, but for the great majority who are willing to live within its means it would be a much more convivial environment for living with others on terms of equal worth, for causes more worthy of our humanity than the infinite competition for earthly goods.

The acceptance of limits is a necessity, but from this acceptance can come the recognition that limitation also allows or even enhances some human potentialities, a realization that is somewhat contrary to our normal expectations of what it means to be limited, or at least counter-intuitive, but has long been understood by traditions of wisdom the world over.[36] Among the social potentialities quickened by the recognition of environmental limits, a new stage in the historical progress of equality is the greatest moral good imaginable. It is of course not guaranteed. Limits can also increase strife, competition, and inequality itself. All the more reason to attend to their better potential.

The case for the good of equality within limits is perhaps even stronger, if no less challenging, when we move from the question of inequality between individuals and groups within nations to that of inequality between nations themselves. As Emmanuel Todd has devoted his lifelong work to explaining, different cultures throughout the world manifest over long periods of time markedly different dispositions or degrees of tolerance toward the fundamental equality or inequality they perceive between human beings, as well as between men and women. There are, in this anthropological sense, more or less egalitarian cultures, and in practice, more or less equal societies.[37] It is of more than speculative interest to wonder whether, in keeping with the preceding argument, more egalitarian countries will prove more hospitable to the idea of shared limits, or if on the contrary more inegalitarian societies will find it harder to accept the moral consequences of a stationary state.[38]

We also know, because the whole of human history provides infinite evidence for it, that different cultures and nations manifest different forms and degrees of openness, respect, or admiration toward other cultures and nations, or on the contrary varying levels of xenophobia, racial or ethnic intolerance, and fear; in most cases in a complex and conflicted combination of all of the above, although it is probable that few have always been immune to collective sentiments of superiority. But in spite of the millennial legacy of conquest, imperialism, and colonialism, no self-conscious people has ever durably accepted its own inferiority. Nor is this any more likely to happen in a world where all peoples, apart from occasional implosions into tribalism and ethnic purification, are organized in nation-states all equally committed to the glorification of their imagined communities and historical narratives, and at the very least expecting equal recognition and rights, to which they are officially entitled by international institutions. Of course, this has thus far not prevented the perpetuation, and in some cases the worsening, of immense inequalities in the real wealth of nations, but in principle, and whatever private prejudices we may still harbor, there is no more official countenance for the essential inequality of national, racial or ethnic collectivities.

This is the context in which to pose the question of global limits. In the international arena, the alibi of inequality is called development. The poorer countries, the majority of which are still dreadfully behind in relation to the richest, are supposed to develop in the material sense, usually using the technological and economic means (financing is a sore point) that have proved successful for the First World, until, presumably, they will catch up at some indefinite future point. And some, most famously in South-East Asia, but also, more haltingly and unevenly, in parts of South America and Africa, are indeed making progress, though they are still long distances away – the only non-Western country to have truly joined the first rank of economies being so far Japan. Even in this indefinite growth model, it is not unreasonable to ask how the runners-up – not to mention the ones that have barely started the race – are in fact going to catch up as long as the leaders are still committed to growth, as they certainly are. Unless there are substantially and durably differential growth rates, the point of equalization must be infinitely far away.[39] An unattainable illusion, an alibi. But now, in part because some of the most populous nations in the world are really developing at record speed and making commensurate demands, and certainly because the most affluent continue to claim a disproportionate share of the world's resources, even the illusion of infinity is finally dissipating. Suddenly, it seems (there were warnings, but they were not heeded), we have reached the limits of the world. Lest we all overrun the unseen finish line, and into the abyss, we must stop the race. Or to use a more homely metaphor, the pie can no longer grow (and the rising tide surely cannot lift all boats). What, then, shall we tell the losers, the ones who are still far behind, the ones with the smaller shares, the ones at the bottom?

In the real world, without metaphors, this is precisely the question being faced in the first major confrontation with limits to reach international prominence, the Global Warming negotiations over allowable levels of greenhouse gas emissions. Since we are unquestionably in overshoot (overheating the atmosphere, which is the very essence of global warming) – again the first such official recognition that we are trespassing planetary limits – it cannot be sufficient to merely stabilize emissions at their present level (a steady state or stationary objective with regards to a particular limit). It is therefore in principle agreed that we must reduce them substantially, a stated goal of global *décroissance*, surely the most ambitious ever undertaken.

But who shall bear the brunt of this reduction, knowing that greenhouse gases are the by-product of fossil fuel burning, still the principal engine of development?[40] An evenly proportional cut of all present emissions would surely be unfair, since the poorer countries, which are hardly responsible for global warming at all, and may well be its first victims, would thereby be practically prevented from any possibility of catching up, while those largely at fault, which in addition have already reaped the greatest benefits, will be better prepared technologically and financially to cope with its results, and suffer the least. A worldwide allocation to each country (an emissions budget, possibly tradeable) based on its respective population seems much more equitable, and even while reducing global emissions to a sustainable level, it would still grant underdeveloped nations much room to grow. But in this case the cuts demanded – with much greater justice – of the largest polluters and beneficiaries would become enormous, much more than is presently contemplated and deemed acceptable by them. And yet such an apportionment would still be far from redressing historical inequalities. The advanced countries have been profiting from the burning of fossil fuels, and spewing greenhouse effluents in the atmosphere – which are for the most part still there, contributing to global warming – for more than 200 years, while the rest of the world was barely making a dent. To achieve real parity, the accounting of emissions and allocation of credits should go back to the industrial revolution.[41] Then justice would indeed be done, and developing countries would have plenty of latitude to catch up. But the rich beneficiaries of the past would have almost nothing left to go on. Before they have time to adapt to renewable energies, they would suffocate from want of dirty air.

So of course it won't happen. In this as in many other cases the wrongs and injustices of history can never be fully redressed, even when they are fully acknowledged. But aside from helping us understand why the negotiations over global warming are so long and hard, and often fail before they succeed in some way, this fable which is not a fable reveals how the acceptance of limits forces us to recognize that we, humans of all the nations, share the earth as never before, and must collectively decide how to do so. For most of history, it mattered little what one people did in their corner of the world until they, or their goods, came to your corner of the world. Now we know that in our full sphere there are no corners any more, and that walls great and small are futile. What one extracts and consumes at one end of our finite planet determines what will be left for another to use; what one discharges in one's river or one's air will eventually affect the whole of the polluted oceans and everyone's saturated sky. Some limits are still local or regional, like available water, or the extent of forest that can be turned into arable land, but in the end every river runs into the sea (well, almost, some are now exhausted long before they can reach their destination), and every bush that is burnt will no longer contribute to the earth's breath, and to our own breathing of the same air.

We are finally one, not only through the warming atmosphere we all breathe and the currents that reach all shores, but also through the finite resources we collectively use up and the goods we trade, and the populations, the ideas, the plants, and the germs we exchange from one end of the earth to every other. Nations not so much united as bound to one another by our common finitude. This new sense of limits, alas, does not by itself guarantee that there will be more harmony, justice and equality among us. Scarcity can on the contrary trigger competition, strife, and violence. Faced with the finality of the goods to be divided between us, humanity will no doubt be tempted to the recourse that has traditionally been its default throughout history, the exercise of power and war, and indeed this is already happening, as few of us are under any illusion that the conflicts in the Middle East are not at least in part wars over oil, with more to come over other finite resources. But when we get tired of the fighting that can only bring us, collectively, closer to the brink, the contemplation of our joint finitude may induce us to seize the ultimate chance, and the greater good, it offers us. The chance to practice, to borrow from the name of a valiant organization of environmental activists, Earth Justice,[42] which can only mean justice among humans. And perhaps for humanity to take, with the acceptance of finality, the first steps toward wisdom.

Notes

1 This is the virtue known to the classical Greeks as *phronesis*, expounded by Aristotle in *The Nicomachean Ethics*, Book VI. Following the work of Cornelius Castoriadis, Jeff Klooger also insists on the value of *phronein* as the exercise of "prudence/wisdom/balance," the cultivation of self-limitation in a democratic *polis* that is no longer imposing absolute norms on the individual (Klooger, 2012, 4–10).

2 In such books as *Impasse Adam Smith* (Michéa, 2002) and *Realm of Lesser Evil* (Michéa, 2009).

3 What characterizes affluent societies, besides the attainment of uniquely high rates of economic growth, is the "revolution of expectations" which is interwoven with the endless creation of new wants that can only be satisfied by ever larger amounts of material wealth.

(Zolotas, 1981, 10)

The United States spent more than three percent of its GNP on advertising for most years from the 1900s to 1930s, and more than two percent since then [more than twice as much as any other advanced industrialized country except for Britain …]. The percentage of national income spent just on advertising in the United States in 1971 was about one-quarter of the outlay on education at all levels.

(Offer, 2006, 122)

4 Perhaps the most telling point of all is the rather too perfect fit between the continual consumption of novelty by households and the continuous production of novelty in firms. The restless desire of the "empty self" is the perfect complement for the restless innovation of the entrepreneur.

(Jackson, 2009, 101)

5 Referring to Vilfredo Pareto's *Manuel d'économie politique* (Pareto, 1927, 18), Georgescu-Roegen writes: "The problem of delimiting the sphere of economics, even in a rough way, is [...] not as simple as Pareto urges us to believe through his argument that just as geometry ignores chemistry so can economics ignore by abstraction *homo ethicus, homo religiosus*, and all other *homines*." And he comments later: "Indeed, if man's economic actions were independent of his cultural propensities, there would be no way to account for the immense variability of the economic pattern with time and locality" (Georgescu-Roegen, 1971, 318, 342).

6 What Tim Jackson calls "the wisdom of the sages from time immemorial." And he adds, in a slightly tangled formulation that seems to reveal that it has, in fact, been diluted: "It hasn't weakened over the years. It hasn't been diluted by our material wealth. If anything, it has simply become harder and harder to see where the real wealth lies" (Jackson, 2009, 190).

7 See Craig Lambert: *The Science of Happiness* (Lambert, 2007).

8 Robert K. Musil gives us an impressive list of individuals and faith-based organizations active in the environmental movement (Musil, 2009, 111–119). See also Lynn White: "The Historical Roots of Our Ecological Crisis" (White, 1967). Of Lynn White's avocation, Donald Worster writes:

> That Lynn White's study was frequently mixed with his religious beliefs is not incidental. He was drawn to an ecological perspective both by an intense attachment to the land and the creatures he had known from childhood and by an equally deep reverence for the divine providence that had contrived this beautiful living whole.
>
> (Worster, 1985, 7)

9 Donald Worster distinguishes usefully between Christian Pastoralism and the Arcadian tradition:

> In the Christian version of the pastoral dream, the shepherd does not merge with nature through his flock nor is his occupation a protest against urban alienation from the natural world, both of which are key themes in the Arcadian version. On the contrary, he is the defender of the flock *against* the hostile forces of nature – wolves, lions, bears – and his profession is to lead his lambs out of this sorry world to greener pastures.
>
> (Ibid., 26)

10 See Kent Flannery and Joyce Marcus: *The Creation of Inequality* (Flannery and Marcus, 2012).

11 Brian Czech argues that all the world's major religions, including the monotheisms, "have sustainable traditions, at least ideally" (Czech, 2013, 152–153).

12 In such books as Ruth Benedict's *Patterns of Culture* (Benedict, 2005) and Margaret Mead's *Coming of Age in Samoa* (Mead, 1928) or her reflections in her autobiographical *Blackberry Winter* (Mead, 1972).

13 In such books as *La Diversité du monde: famille et modernité*, translated as *The Explanation of Ideology: Family Structure and Social Systems* (Todd, 1983; 1989), and *L'Origine des systèmes familiaux* (Todd, 2011), and many essays in applied sociology.

14 See Dana L. and Laura L. Jackson, *The Farm as Natural Habitat* (Jackson and Jackson, 2002).

15 It is appropriate to remember that wilderness – and here wildlife itself – are in Fred Hirsch's terms "positional goods," their value is a function of their rarity, and diminishes, or is degraded, as they become accessible to greater numbers of people (Hirsch, 1976, 21). This was already understood by Aldo Leopold with his even more relevant notion of "trophies" (Leopold, 1949, 169–172).

16 In addition to the damage of overgrazing by those that are allowed some degree of free range, and the immense resource drain, soil erosion and pollution caused by the

soy and corn and other monocultures used to fatten those in feedlots – livestock eat 70 percent of all corn, wheat and grain produced in the US (Musil, 2009, 194) – let us remember that cattle are responsible for 15 percent of greenhouse gas emissions worldwide, and growing (Hawken, 2017, 39).

17 Among the many accessible documents and reflections on the instrumental use of animals, see the American Society for the Prevention of Cruelty to Animals article on "Farm Animals Need Our Help" (ASPCA, n.d.) and Michael Pollan's exposition on the same subject: "An Animal's Place" (Pollan, 2002). For a thoughtful discussion of the use of animals for research benefiting humans, see the US National Research Council's *Recognition and Alleviation of Pain in Laboratory Animals* (National Research Council, 2009).

18 *An Introduction to the Principles of Morals and Legislation* (Bentham, 1996, 283). Peter Singer quotes and comments on it in *Practical Ethics* (Singer, 1980, Chapter 3). In "The Surprisingly Humanlike Ways Animals Feel Pain," Liz Langley confirms: "Mammals share the same nervous system, neurochemicals, perceptions, and emotions, all of which are integrated into the experience of pain." Mammals are most like humans in their ability to feel and express pain; birds feel pain akin to the pain mammals feel; reptiles, amphibians, and fish also can feel pain although it is more different due to their neuroanatomy (Langley, 2016).

19 See in particular the debates about animal sensitivity in Erica Harth: *Cartesian Women* (Harth, 1992).

20 See *Animal Rights and Human Obligations* (Regan and Singer, 1976, 59; 122), discussed by Martin Benjamin in "Ethics, Animals, and Language" (De Luce and Wilder (eds.), 1983, 113–115).

21 See Stephen T. Newmyer: "Plutarch on the Moral Grounds of Vegetarianism" (Newmyer, 1995).

22 Even though meat consumption is steadily increasing in China and India and other fast developing countries, "'the average Indian consumes a thirtieth of the meat of an Australian or an American does – around 4.4 kg in 2009' whereas in the US it is '120 kg per head per annum, as much or more meat than anyone.'" The UK is not very far behind, at 89.1 kg a year. From Bee Wilson: "How Much Meat is too Much," quoting from Alex Renton's *Planet Carnivore* (Wilson, 2014; Renton, 2013).

23 See among many studies Lisa-Ann Gershwin: *Stung! On Jellyfish Blooms and the Future of the Ocean* (Gershwin, 2013).

24 As reported by Renee Cho, "The Food and Agriculture Organization of the U.N. (FAO) projects that by 2030, aquaculture [...] will be responsible for almost two thirds of the fish we eat." "Over 50 percent of the world's fish oil is used in feed for farmed salmon," as it takes "almost three tons of forage fish to produce one ton of salmon." In addition,

> Fish waste and leftover food spill out from nets into the ocean, causing nutrient pollution, [...] antibiotics or pesticides used on farmed fish can affect other marine life or human health. [...] Fish crowded together in nets or pens are more susceptible to stress, which can foster disease and parasites that may then spread to wild species. Farmed fish sometimes escape into the ocean, breeding with wild species and affecting the population's overall genetic diversity.
>
> (Cho, 2016)

25 Again, as reported by Karen Higgins, and consistent with note 17 (this chapter), livestock methane was responsible for 14 to 18 percent of worldwide greenhouse gas emissions in 2006 (Higgins, 2015, 57).

26 It should be noted, however, that Monterey Bay Aquarium's Seafood Watch researchers maintain that these negative effects could be minimized, and that aquaculture "holds great promise as a solution to the increasing pressures on our ocean resources." With good practices, they believe, "it's possible to farm seafood in a way

that has very little impact on the environment. Such operations limit habitat damage, disease, escapes of farmed fish and the use of wild fish as feed" (Seafood Watch, n.d.).

27 See David Pimentel and Marcia Pimentel: "Sustainability of Meat-Based and Plant-Based Diets and the Environment" (Pimentel and Pimentel, 2003).

28 According to James Hansen, the "current extinction rate is at least 100 times greater than the average natural rate," and "several thousand species are becoming extinct every year" (Hansen, 2009, 147). But as we have already noted, Thomas L. Friedman reports that "Conservation International currently estimates that one species is now going extinct every 20 minutes, which is 1,000 times faster than the norm during most of the earth's history" (Friedman, 2008, 141). Such a wide discrepancy between estimates is partly explained by the fact that we do not how many species exist, and we will never know many of them before they disappear.

29 Again in the words of Bill McKibben already quoted: "We're running Genesis backward, decreating" (McKibben, 2010, 206).

30 In 2017 The United Nations Department of Economic and Social Affairs projected a global population of 9.8 billion in 2050, and 11.2 billion in 2100 (UN DESA, 2017a). Another UN report from 2017, however, put the estimate for 2100 between a possible low of 9.8 billion and as high as 13.2 billion (UN DESA, 2017b).

31 See the invaluable study *The Meaning of Things: Domestic Symbols and the Self*, by Mihali Csikszentmihalyi and Eugene Rochberg-Halton (Csikszentmihalyi and Rochberg-Halton, 1981).

32 As we saw in the previous chapter, some of the most informative and possibly hopeful answers to these questions are notably suggested by Richard Wilkinson and Kate Pickett in *The Spirit Level* (Wilkinson and Pickett, 2009), and by Fred Hirsch in *Social Limits to Growth* (Hirsch, 1976).

33 See again Avner Offer, *The Challenge of Affluence*, particularly pp. 31–32 and 274–275 (Offer, 2006), as well as the books praised in the previous note.

34 Recalling the Smithian understanding of "sympathy," but we could equally evoke the Hegelian "struggle for recognition" emphasized by Arran E. Gare (Gare, 2000), Avner Offer argues that well-being is achieved in the "economy of regard": "Under reciprocity, the regard of others is earned by providing them with our own regard, directly or indirectly" (Offer, 2006, 267).

35 And again the compelling data provided by Richard Wilkinson and Kate Pickett in *The Spirit Level* (Wilkinson and Pickett, 2009), as well as in Oliver James's *Affluenza* (James, 2007). On the basis of a presentation at the University of Surrey, Tim Jackson writes similarly: "Tim Kasser and his colleagues show that people in liberalized market economies tend to have higher per capita carbon emissions, higher infant mortality, higher teenage pregnancies and a greater percentage of people reporting that they 'feel like an outsider'" (Jackson, 2009, 165).

36 At the end of their own investigation of our investment in (cultural) objects, Mihaly Csikszentmihalyi and Eugene Rochberg-Halton see the escape from "terminal materialism" in the cultivation of new meanings:

> There is an urgent need to develop vital cultural practices – ways of living – that have the potential for *long-term* growth. We need to cultivate those objects, activities, and environments that can inspire the fullest unfolding of human potential now and in the future; patterns of meaning that do not have a self-destructive exhaustion of physical and psychic resources built into them.
>
> (Csikszentmihalyi and Rochberg-Halton, 1981, 248)

37 There are, objectively speaking, vast differences between levels of inequality within countries, as measured by their GINI coefficient. But as Joseph Stiglitz points out, "countries differ not just in how unequal they are now but also in their culture, tolerance of inequality of various kinds, and capacity for social change" (Stiglitz, 2015, 292).

38 Significantly, Naomi Klein, summarizing the findings of the Yale Cultural Cognition Project, reports that:

> The Yale researchers explain that people with strong "egalitarian" and "communitarian" world views (marked by an inclination toward collective action and social justice, concern about inequality, and suspicion of corporate power) overwhelmingly accept the scientific consensus on climate change. Conversely, those with strong "hierarchical" and "individualistic" worldviews (marked by opposition to government assistance for the poor and minorities, strong support of industry, and a belief that we all pretty much get what we deserve) overwhelmingly reject the scientific consensus.
>
> (Klein, 2014, 36)

39 See the troubling conclusions of Branko Milanovic in *Worlds Apart: Measuring International and Global Inequality*, especially Chapters 11 and 12 (Milanovic, 2005).

40 For an overview of the philosophical issues, see Stephen M. Gardiner: "Ethics and Climate Change: An Introduction" (Gardiner, 2010).

41 The "legacy" ranking of countries by totaling their cumulative emissions through the modern historical period is quite different from the ranking by current levels of emissions. In the cumulative record, the order is (1) UK, (2) US, (3) Germany, (4) Canada, (5) Japan, (6) China, and (7) India (Hansen, 2009, 177). According to the World Resources Institute, the top emitter nations in 2017 were (1) China, (2) US, (3) EU (28 countries), (4) India, (5) Russia, (6) Japan, (7) Brazil (www.wri.org/blog/2017/04/interactive-chart-explains-worlds-top-10-emitters-and-how-theyve-changed). For the per capita ranking, three years earlier, the same organization gives us: (1) Canada, (2) US, (3) Russia, (4) Japan, (5) EU, (6) Indonesia, (7) China (www.wri.org/blog/2014/11/6-graphs-explain-world's-top-ten-emitters).

42 The Earth Justice website can be found at https://earthjustice.org/.

References

Aristotle (2003) *The Nicomachean Ethics*. London: Penguin Classics.

ASPCA (n.d.) "Farm Animals Need Our Help" [online]. Available at www.aspca.org/animal-cruelty/farm-animal-welfare.

Benedict, Ruth (2005) *Patterns of Culture*. Boston and New York: Mariner Books.

Bentham, Jeremy (1996) *An Introduction to The Principles of Morals and Legislation*, J. H. Burns and H. L. A. Hart (eds.). Oxford: Oxford University Press.

Cho (2016) "Making Fish Farming More Sustainable" [online]. Available at https://blogs.ei.columbia.edu/2016/04/13/making-fish-farming-more-sustainable/.

Csikszentmihalyi, Mihali and Eugene Rochberg-Halton (1981) *The Meaning of Things: Domestic Symbols and the Self*. Cambridge, UK: Cambridge University Press.

Czech, Brian (2013) *Supply Shock: Economic Growth at the Crossroads and the Steady State Solution*. Gabriola Island, DC, Canada: New Society Publishers.

De Luce, Judith and High Wilder (eds.) (1983) *Language in Primates: Perspectives and Implications*. New York: Springer-Verlag.

Dyson, Freeman (2008) "The Question of Global Warming," *New York Review of Books*, June 12.

Flannery, Kent and Joyce Marcus (2012) *The Creation of Inequality: How Our Ancestors Set the Stage for Monarchy, Slavery, and Empire*. Cambridge, MA: Harvard University Press.

Friedman, Thomas L. (2008) *Hot, Flat, and Crowded: Why We Need a Green Revolution – and How It Can Renew America*. New York: Farrar, Strauss & Giroux.

Gare, Arran E. (2000) "Is It Possible to Create an Ecologically Sustainable World Order: The Implications of Hierarchy Theory for Human Ecology," *International Journal of Sustainable Development and World Ecology*, vol. 7, no. 4, 277–290.
Gardiner, Stephen M. (2010) "Ethics and Climate Change: An Introduction," *WIRE's Climate Change*, vol. 1, January/February, 54–66.
Garvey, James (2008) *The Ethics of Climate Change: Right and Wrong in a Warming World*. London: Continuum.
Georgescu-Roegen, Nicholas (1971) *The Entropy Law and the Economic Process*. Cambridge, MA and London: Harvard University Press.
Gershwin, Lisa-Ann (2013) *Stung! On Jellyfish Blooms and the Future of the Ocean*. Chicago: Chicago University Press.
Hansen, James (2009) *Storms of my Grandchildren: The Truth about the Coming Climate Catastrophe and our Last Chance to Save Humanity*. New York: Bloomsbury USA.
Harth, Erica (1992) *Cartesian Women: Versions and Subversions of Rational Discourse in the Old Regime*. Ithaca and London: Cornell University Press.
Hawken, Paul (ed.) (2017) *Drawdown: The Most Comprehensive Plan Ever Proposed to Reverse Global Warming*. New York: Penguin Books.
Higgins, Karen L. (2015) *Economic Growth and Sustainability: Systems Thinking for a Complex World*. Academic Press, Elsevier.
Hirsch, Fred (1976) *Social Limits to Growth*. Cambridge, MA: Harvard University Press.
Hirschman, Albert O. (1977) *The Passions and the Interests: Political Arguments for Capitalism before its Triumph*. Princeton, NJ: Princeton University Press.
Jackson, Dana L. and Laura L. Jackson (2002) *The Farm as Natural Habitat: Reconnecting Food Systems with Ecosystems*. Washington: Island Press.
Jackson, Tim (2009) *Prosperity without Growth*. London and New York: Earthscan-Routledge.
James, Oliver (2007) *Affluenza*. London: Vermilion.
Klein, Naomi (2014) *This Changes Everything: Capitalism vs. The Climate*. New York: Simon & Schuster.
Klooger, Jeff (2012) "The Meanings of Autonomy: Project, Self-Limitation, Democracy and Socialism," *Thesis Eleven*, vol. 108, no. 1, 84–85.
Lambert, Craig (2007) "The Science of Happiness," *Harvard Magazine*, January/February.
Langley, Liz (2016) "The Surprisingly Humanlike Ways Animals Feel Pain," National Geographic, December 3, 2016 [online]. Available at https://news.nationalgeographic.com/2016/12/animals-science-medical-pain/).
Leopold, Aldo (1949) *A Sand County Almanac, and Sketches Here and There*. New York and Oxford: Oxford University Press.
Lévi-Strauss, Claude (1997) *Tristes Tropiques*. New York: Washington Square Press.
Manning, Richard (2004) *Against the Grain: How Agriculture Hijacked Civilization*. New York: North Point Press.
McKibben, Bill (2010) *Eaarth: Making a Life on a Tough New Planet*. New York: Holt.
Mead, Margaret (1928) *Coming of Age in Samoa: A Psychological Study of Primitive Youth for Western Civilization*. New York: William Morrow and Company.
Mead, Margaret (1972) *Blackberry Winter: My Earlier Years*. New York: William Morrow and Company.
Merkel, Jim (2003) *Radical Simplicity: Small Footprints on a Finite Earth*. Gabriola Island, BC: New Society Publishers.
Michéa, Jean-Claude (2002) *Impasse Adam Smith*. Castelnau-le-Lez, France: Climats.
Michéa, Jean-Claude (2009) *Realm of Lesser Evil*. Cambridge, UK: Polity Press.

Milanovic, Branko (2005) *Worlds Apart: Measuring International and Global Inequality*. Princeton, NJ: Princeton University Press.

Montgomery, David R. (2007) *Dirt: The Erosion of Civilizations*. Berkeley, CA, Los Angeles and London: University of California Press.

Moore Lappé, Frances (1971) *Diet for a Small Planet*. New York: Ballantine Books.

Musil, Robert K. (2009) *Hope for a Heated Planet: How Americans Are Fighting Global Warming and Building a Better Future*. New Brunswick, NJ and London: Rudgers University Press.

National Research Council (2009) *Recognition and Alleviation of Pain in Laboratory Animals*. Washington, DC: National Academies Press.

Newmyer, Stephen T. (1995) "Plutarch on the Moral Grounds of Vegetarianism," *The Classical Outlook*, vol. 72, no. 2 (Winter).

Offer, Avner (2006) *The Challenge of Affluence: Self-Control and Well-Being in the United States and Britain since 1950*. Oxford: Oxford University Press.

Pareto, Vilfredo (1927) *Manuel d'économie politique*. Paris.

Peterson, Anna L. (2001) *Being Human: Ethics, Environment, and Our Place in the World*. Berkeley, CA and Los Angeles: University of California Press.

Pimentel, David and Marcia Pimentel (2003) "Sustainability of Meat-Based and Plant-Based Diets and the Environment," *The American Journal of Clinical Nutrition*, vol. 78, no. 3, September, 660S–663S.

Pollan, Michael (2002) "An Animal's Place," *New York Times Magazine*, November 10.

Regan, Tom and Peter Singer (eds.) (1976) *Animal Rights and Human Obligations*. Englewood Cliffs, NJ: Prentice Hall.

Renton, Alex (2013) "Planet Carnivore: Why Cheap Meat Costs the Earth (and How to Pay the Bill)" (extract), *Guardian*, September 17.

Schumacher, E. F. (2010) *Small Is Beautiful: Economics as if People Mattered*. New York: Harper Perennial.

Seafood Watch (n.d.) "Aquaculture" [online]. Available at www.seafoodwatch.org/ocean-issues/aquaculture.

Sen, Amartya (1992) *Inequality Reexamined*. Cambridge, MA: Harvard University Press.

Singer, Peter (1980) *Practical Ethics*. New York: Cambridge University Press.

Smith, Adam (1759) *The Theory of Moral Sentiments*. Oxford: Oxford University Press (this edition 1976).

Stiglitz, Joseph E. (2015) *The Great Divide: Unequal Societies and What We Can Do About Them*. New York: Norton.

Todd, Emmanuel (1983) *La Diversité du monde: famille et modernité*. Paris: Seuil.

Todd, Emmanuel (1989) *The Explanation of Ideology: Family Structure and Social Systems*. New York: Blackwell Publishing.

Todd, Emmanuel (2011) *L'Origine des systèmes familiaux*. Paris: Gallimard, 2011.

UN DESA (2017a) "World Population Projected to Reach 9.8 Billion in 2050, and 11.2 Billion in 2100," June 21, 2017 [online]. Available at www.un.org/development/desa/en/news/population/world-population-prospects-2017.html.

UN DESA (2017b) "World Population Prospects 2017" [online]. Available at https://population.un.org/wpp/Graphs/Probabilistic/POP/TOT/.

Weisman, Alan (2007) *The World without Us*. New York: Thomas Dunne Books, St. Martin's Press.

White, Lynn, Jr. (1967) "The Historical Roots of Our Ecological Crisis," *Science*, vol. 155, no. 3767, 1203–1207.

Wilkinson, Richard and Kate Pickett (2009) *The Spirit Level: Why Greater Equality Makes Societies Stronger*. New York: Bloomsbury Press.

Wilson, Bee (2014) "How Much Meat Is Too Much?" *London Review of Books*, March 20.

Worster, Donald (1985) *Nature's Economy: A History of Ecological Ideas*. Cambridge, MA: Cambridge University Press.

Zolotas, Xenophon (1981) *Economic Growth and Declining Social Welfare*. New York and London: New York University Press.

Index

For Product Safety Concerns and Information please contact our EU representative GPSR@taylorandfrancis.com Taylor & Francis Verlag GmbH, Kaufingerstraße 24, 80331 München, Germany